Nafees A. Khan (Ed.)

Ethylene Action in Plants

With 31 Figures, 1 in Color, and 6 Tables

 Springer

Dr. Nafees A. Khan
Department of Botany
Aligarh Muslim University
Aligarh 202002
India
e-mail: naf9@lycos.com

Library of Congress Control Number: 2006923693

ISBN-10 3-540-32716-9 Springer Berlin Heidelberg New York
ISBN-13 978-3-540-32716-5 Springer Berlin Heidelberg New York

This work is subject to copyright. All rights are reserved, whether the whole or part of the material is concerned, specifically the rights of translation, reprinting, reuse of illustrations, recitation, broadcasting, reproduction on microfilm or in any other way, and storage in data banks. Duplication of this publication or parts thereof is permitted only under the provisions of the German Copyright Law of September 9, 1965, in its current version, and permissions for use must always be obtained from Springer-Verlag. Violations are liable for prosecution under the German Copyright Law.

Springer is a part of Springer Science+Business Media

springer.com

© Springer-Verlag Berlin Heidelberg 2006
Printed in the Netherlands

The use of general descriptive names, registered names, trademarks, etc. in this publication does not imply, even in the absence of a specific statement, that such names are exempt from the relevant protective laws and regulations and therefore free for general use.

Cover design: Design & Production GmbH, Heidelberg, Germany
Typesetting and production: SPi

Printed on acid-free paper   149/3150-YL/SPI   5 4 3 2 1 0

Ethylene Action in Plants

Foreword

The discovery of the plant hormone ethylene was stunning—ethylene is a simple gas! Our expanding knowledge of the multiplicity of ethylene's roles in plant development, physiology, and metabolism makes the study of this plant hormone increasingly compelling. Elucidation of the genetic regulation of ethylene biosynthesis, characterization of ethylene receptors and analysis of the pathway of ethylene signal transduction, coupled with the identification of components in the cascade and target genes, have provided insight into how this simple molecule can drive such a diversity of divergent processes. These scientific advances will lead to new technologies that will further enable researchers to harness the powers of ethylene for the benefit of agriculture.

In *Ethylene Action in Plants*, classic and emerging roles of ethylene in plant developmental processes are integrated through recent advances characterizing ethylene receptors, promoters and antagonists, and biological and environmental factors that mediate ethylene responses. The book's editor, Dr. Nafees Khan, Aligarh Muslim University, Aligarh, India, an expert on ethylene with an impressive number of publications on the interactions between ethylene, photosynthesis, and growth of *Brassica* spp, brought together a highly qualified group of international experts to provide state-of-the-art information. To simply list the topics included does not do justice to the book's contents, as the articles are not just a compilation of the literature relevant to the topic. The authors have synthesized traditional ethylene research with recent novel discoveries to provide both the means for understanding what have previously been considered conflicting results and answers to previously unanswered questions. The book is designed to provide the reader with the details of major strides in ethylene research, including introduction to new areas of research. I offer the following as a brief glimpse into the pages of *Ethylene Action in Plants*.

Ethylene has long been known as the "ripening hormone", but in recent years progress in identifying ethylene receptors in responsive cells and components of the ethylene signal transduction pathway, including transcription factors and target genes controlling ripening-related processes in fruit and vegetables, has been dramatic. In *Ethylene Action in Plants*, advances in the genetic regulation of ripening are detailed in relation to the role played by other hormones and with goal of delineating the differences among developmentally regulated, ethylene early responsive and ethylene late responsive genes. Knowledge of the molecular basis for fruit ripening will undoubtedly result in improved post-harvest longevity, and increased aesthetic and nutritional quality. In recent years, the investigation of promoters and antagonists to the binding of ethylene with its receptor has lead to the identification of numerous compounds that mediate the interaction. The binding and activity of these compounds are described, along with their potential benefit to basic research and agricultural. It is the hope that the specifics given in the book might lead its readers to discover additional regulatory compounds of value. Whereas the role of ethylene in expansion growth is well known, its effects on biomass accumulation remain understudied, particularly in relation to plants growing under limiting environmental conditions, where ethylene should logically be a factor in the growth response of the plant. In *Ethylene Action in Plants*, the effects of endogenous and exogenous ethylene on growth parameters in optimal and stressful environments are unraveled. Enhanced ethylene production is a common plant response to numerous stresses, but recent evidence that ethylene perception and signal transduction are also affected by stress has lead to the new insight into ethylene sensitivity during stress and stress adaptation presented in the book. Leaf senescence, the last phase of leaf development, is a genetically programmed process. Ethylene plays a key role among leaf senescence inducers. In *Ethylene Action in Plants*, the sequence of events resulting in leaf senescence is described in detail in relation to the physiological effects of ethylene on the process and in light of new research on the modification of ethylene effects by biological and environmental factors that act as promoters and antagonists of ethylene. At best, the role of ethylene in adventitious root development is confusing due the variable responses to ethylene reported in the literature. These variable responses are discussed with the outcome being a better understanding of the basis for the variability and resolution of the conflict. Ethylene also mediates nodulation responses of roots. Comparison of different rhizobium-legume symbioses and their respective nodulation processes provides clarification of contrasting requirements for ethylene in the different bacterial invasion mechanisms involved in nodulation and of the role of ethylene in further nodule development. Another controversial aspect of ethylene physiology discussed in the book is ethylene's role in regulating stem gravitropic curvature. Here, historic evidence and traditional methods are critically evaluated in light of recent advances in the field. The interactions between ethylene and photosynthesis and growth are complex due to modulation of the effects of ethylene by many factors. Evidence is provided to support the involvement of 1-aminocyclopropane carboxylic acid synthase, the rate-limiting enzyme in the synthesis of ethylene, as a common factor in the control of photosynthesis and growth. *Ethylene Action in Plants* integrates results from physiology, biochemistry, and molecular biology research.

Readers of *Ethylene Action in Plants* will gain an appreciation for how significantly our understanding of ethylene action has advanced in recent years and for current efforts by researchers to answer those questions that remain unanswered and to pose new questions. The book will expand the knowledge base and stimulate the thinking of plant biology graduate students and researchers, be they botanists, ecologists, horticulturists, agronomists, physiologists, molecular biologists, or genetic engineers.

Carol J. Lovatt, Ph.D.,
Professor of Plant Physiology
University of California -Riverside

Preface

Ethylene, the simplest plant growth regulator, has been recognized to control many physiological processes in plants, including fruit ripening, abscission, senescence, and responses of biotic and abiotic stresses. Since the time of the Egyptians, ethylene has been used to stimulate the ripening of figs and the Chinese used it to enhance the ripening of pears. The phenomenon of 'triple response' induced by ethylene was discovered in 1864 when it was noticed that gas leaks from street lamps caused stunting of growth, twisting of plants, and abnormal thickening of stems. It was Neljubow in 1901 who discovered that the active principle in illuminating gas was ethylene; thus, he is credited with the discovery that ethylene is a biologically active gas. Later, in 1934, Gane provided chemical proof that plants produce ethylene. It has now been recognized that ethylene is produced in all higher plants. Thus, with the recognition of the presence of ethylene in plants, the stage was set to investigate the ethylene action in plants as signal molecules. The action of ethylene as a signal molecule depends on its tissue concentration and the ability of the cells to monitor the changing concentrations of ethylene and transduce this information into physiological responses. The effectiveness of ethylene requires high-affinity receptors. It is bound by a membrane-localized receptor. The N-terminal domain of the receptor protein is responsible for binding of ethylene. Components of the ethylene signal transduction pathway have been identified by genetic studies in *Arabidopsis*. Transduction of the ethylene signal is thought to be achieved through a series of phosphorylations that are carried out by a cascade of protein kinases similar to the mitogen-activated protein kinase pathway. Genetic manipulation of the genes responsible for the ethylene signal transduction pathway will provide agriculture with new tools to prevent or modify ethylene responsible in a variety of plants.

    The intent of this book is not to cover all the aspects of ethylene biology but to summarize and provide an update on our current understanding on mechanism and regulation of ethylene action. I extend my gratitude to all those who have contributed in making this book possible. Simultaneously, I would like to apologize unreservedly for any mistakes or failure to acknowledge fully. Finally, I thank my family for their continued support and encouragement throughout the work.

List of Contributors

DEN HERDER, JEROEN
Department of Plant Systems Biology, Flanders Interuniversity Institute for Biotechnology (VIB), Ghent University, Technologiepark 927, B-9052, Gent, Belgium

DRUEGE, UWE
Institute of Vegetable and Ornamental Crops, Grossbeeren/Erfurt e.V. Department Plant Propagation, Kuehnhaeuser Str., 101, 99189 Erfurt-Kuehnhausen, Germany

FERRANTE, ANTONIO
Dep. Produzione Vegetale, University of Milan, via Celoria 2, I-20133, Italy

FRANCINI, ALESSANDRA
Dep. Coltivazione e Difesa delle Specie Legnose 'G. Scaramuzzi' University of Pisa via del Borghetto 80, I-56100, Italy

GOORMACHTIG, SOFIE
Department of Plant Systems Biology, Flanders Interuniversity Institute for Biotechnology (VIB), Ghent University, Technologiepark 927, B-9052, Gent, Belgium

GRICHKO, VARVARA P.
Department of Molecular and Structural Biochemistry, North Carolina State University, Campus Box 7622, Raleigh, NC 27695, USA

HARRISON, MARCIA A.
Department of Biological Sciences, Marshall University, Huntington, WV 25755, USA

HOLSTERS, MARCELLE
Department of Plant Systems Biology, Flanders Interuniversity Institute for Biotechnology (VIB), Ghent University, Technologiepark 927, B-9052, Gent, Belgium

KHAN, NAFEES A.
Department of Botany, Aligarh Muslim University, Aligarh 202002, India

NATH, PRAVENDRA
Plant Gene Expression Laboratory, National Botanical Research Institute, Rana Pratap Marg, Lucknow 226001, India

PAN, RUICHI
College of Life Science, South China Normal University, Guangzhou 510631, China

POORTER, HENDRIK
Utrecht University, Sorbonnelaan 16, Utrecht, 3584 CA, The Netherlands

SANE, ANIRUDDHA P.
Plant Gene Expression Laboratory, National Botanical Research Institute, Rana Pratap Marg, Lucknow 226001, India

SANE, VIDHU A.
Plant Gene Expression Laboratory, National Botanical Research Institute, Rana Pratap Marg, Lucknow 226001, India

SEREK, MARGRETHE
Department of Agricultural Sciences, Crop Science, Royal Veterinary and Agricultural University, Thorvaldsensvej 40, 1871, Frederiksberg C, Denmark

SISLER, EDWARD C.
Department of Molecular and Structural Biochemistry, North Carolina State University, Campus Box 7622, Raleigh, NC 27695, USA

THOLEN, DANNY
Utrecht University, Sorbonnelaan 16, Utrecht, 3584 CA, The Netherlands

TRIVEDI, PRABODH K.
Plant Gene Expression Laboratory, National Botanical Research Institute, Rana Pratap Marg, Lucknow 226001, India

VOESENEK, LAURENTIUS A.C.J.
Utrecht University, Sorbonnelaan 16, Utrecht, 3584 CA, The Netherlands

WANG, JINXIANG
College of Resources and Environment and Root Biology Centre, South China Agricultural University, Guangzhou 510642, China

Contents

1. **Interaction of Ethylene and Other Compounds with the Ethylene Receptor: Agonists and Antagonists**
   EDWARD C. SISLER, VARVARA P. GRICHKO AND MARGRETHE SEREK 1
   1.1 Introduction 1
   1.2 Ethylene and Agonists... 1
       1.2.1 Discovery of Ethylene Action and Some Important Lessons from the Past 1
       1.2.2 Molecular Requirements................................. 2
       1.2.3 Ethylene Binding 4
   1.3 Ethylene Antagonists.. 5
       1.3.1 Chemicals Adjuvants Counteracting Ethylene 5
       1.3.2 Ethylene Agonists That Require Continuous Exposure to Give a Response..................................... 6
       1.3.3 Naturally Occurring Ethylene Antagonists 9
       1.3.4 Photoactivated Compounds 9
   1.4 Ethylene Agonists That Require a Single Exposure 12
       1.4.1 Ring Strain and Cyclopropene 12
       1.4.2 1-Methylcyclopropene................................... 13
       1.4.3 3-Methylcyclopropene................................... 13
       1.4.4 3,3-Dimethylcyclopropene 15
       1.4.5 Other 1-Alkyl Cyclopropenes 15
       1.4.6 Other Cyclopropenes.................................... 15
       1.4.7 Is the Binding Site Restricted? 16
       1.4.8 Hydrophobic Interaction 17
   1.5 How Does an Agonist Start a Signal?........................... 18
   1.6 Selected Papers on Significance of Ethylene Antagonists 19
       1.6.1 Application in Molecular Biology....................... 19
       1.6.2 Application for Control of Plant Growth and Development 22
       1.6.3 Commercial Use... 24
       1.6.4 Limitations of Compounds............................... 26
       1.6.5 Future Needs 28
   1.7 Concluding Remarks.. 28
       References 29

2. Ethylene and Plant Growth
DANNY THOLEN, HENDRIK POORTER AND LAURENTIUS A.C.J. VOESENEK 35
- 2.1 Introduction 35
- 2.2 Plant Growth Rate 36
  - 2.2.1 Plant Growth Analysis 37
- 2.3 Ethylene and Leaf Expansion 38
- 2.4 Ethylene and Photosynthesis 41
- 2.5 Ethylene and Growth under Optimal Conditions 42
- 2.6 Ethylene and Growth under Limiting Environmental Conditions 43
- 2.7 Ethylene: Effects on Plant Growth? 45
- References 46

3. Ethylene and Leaf Senescence
ANTONIO FERRANTE AND ALESSANDRA FRANCINI 51
- 3.1 Introduction 51
- 3.2 Ethylene during Leaf Development 52
- 3.3 Ethylene-Induced Leaf Senescence 53
  - 3.3.1 Leaf Yellowing 53
  - 3.3.2 Leaf Abscission 54
- 3.4 Effect of Ethylene Inhibitors, Promoters, and Ethylene-Releasing Compounds on Leaf Senescence ... 55
- 3.5 Leaf Sensitivity and Ethylene Production 58
- 3.6 Effect of Ethylene on the Antioxidant System during Leaf Senescence 59
- 3.7 Ethylene and Other Plant Hormones during Leaf Senescence 61
- 3.8 Ethylene and Gene Expression during Leaf Senescence ... 61
- References 64

4. Effect of Ethylene on Adventitious Root Formation
JINXIANG WANG AND RUICHI PAN 69
- 4.1 Introduction 69
- 4.2 Ethylene Stimulates Adventitious Root Formation ... 70
- 4.3 Ethylene Inhibits Adventitious Rooting 72
- 4.4 Ethylene Has No Role in Adventitious Root Formation ... 74
- 4.5 Interactions between Ethylene and Other Hormones ... 74
- 4.6 Discussion 75
- 4.7 Outlook 77
- References 77

5. Ethylene and Plant Responses to Abiotic Stress
UWE DRUEGE 81
- 5.1 Introduction 81
- 5.2 Abiotic Stress and Ethylene Biosynthesis 81
  - 5.2.1 Wounding and Mechanical Stress 82
    - 5.2.1.1 Wound Ethylene in Fruits 84

		5.2.1.2	Wound Ethylene in Vegetative Tissues	86

		5.2.1.2 Wound Ethylene in Vegetative Tissues	86
		5.2.1.3 Mechanical Non-Injury Stress	88
	5.2.2	Water Deficit Stress	89
	5.2.3	Salinity	91
	5.2.4	Flooding/Hypoxia	93
	5.2.5	Chilling	95
	5.2.6	Ozone	97
5.3	Stress-Mediated Ethylene Sensitivity		98
5.4	Ethylene Action, Adaptation and Stress Tolerance		102
	5.4.1	Wound Response and Thigmomorphogenesis	103
	5.4.2	Ethylene and Ozone-Induced Cell Death	103
	5.4.3	Ethylene and Root Stress-Mediated Shoot Growth	106
	5.4.4	Ethylene and Adaptation to Salt Stress	107
	5.4.5	Ethylene and Flooding Tolerance	109
	5.4.6	Ethylene and Chilling Response	109
5.5	Concluding Remarks		110
	References		111

6. Ethylene in the *Rhizobium*-Legume Symbiosis
Jeroen Den Herder, Sofie Goormachtig and Marcelle Holsters 119

6.1	An Introduction to Legume Nodulation		119
6.2	Ethylene-Sensitive RHC Nodulation		121
	6.2.1	Pharmacological Evidence	121
	6.2.2	Mutant Analysis and Transgenic Approaches in Plants and Bacteria	123
	6.2.3	Ethylene Interferes with NF Signaling Within the Root Hairs	125
6.3	Ethylene is Indispensable for LRB Nodulation		125
	6.3.1	Pharmacological Data Show That Ethylene is Needed for Crack-Entry Invasion	126
	6.3.2	Ethylene Mediates the Phenotypic Plasticity in Root Nodule Development	126
	6.3.3	Ethylene Mediates the Switch from Intercellular to Intracellular Invasion	127
6.4	Ethylene Determines Nodule Primordium Positioning		128
6.5	Long-Distance Regulation Does Not Involve Ethylene		129
6.6	Conclusions		129
	References		130

7. The Role of Ethylene in the Regulation of Stem Gravitropic Curvature
Marcia A. Harrison 135

7.1	The Modulating Role of Ethylene on Stem Gravitropic Curvature	135
7.2	Overview of Curvature Kinetics for Light-Grown Compared to Dark-Grown Tissues	136
7.3	Ethylene Production Increases after Horizontal Reorientation and during Curvature	138

7.4	\multicolumn{2}{l}{Differing Sensitivities to Exogenous Ethylene may alter the Gravitropic Response ....}	140

7.4 Differing Sensitivities to Exogenous Ethylene may alter
   the Gravitropic Response 140
   7.4.1 Evidence of a Stimulatory Role for Ethylene in
          Stem Gravitropism.................................. 140
   7.4.2 Ethylene's Role in Slowing Gravitropic Curvature 143
7.5 Ethylene and Auxin Cross-Talk in the Regulation
   of Gravitropic Curvature................................... 144
7.6 Concluding Remarks.. 147
   References 148

8. Role of Ethylene in Fruit Ripening
Pravendra Nath, Prabodh K. Trivedi, Vidhu A. Sane and
Anirudha P. Sane 151
8.1 Introduction... 151
8.2 What is Fruit Ripening? 152
8.3 Ethylene Biosynthesis in Fruits 154
   8.3.1 ACC Synthase in Fruits 154
   8.3.2 ACC Oxidase in Fruits................................ 155
8.4 Ethylene Perception and Signal Transduction in Fruits 156
8.5 Gene Expression and Regulation during Ripening in Fruits 158
   8.5.1 Developmental Regulation 159
   8.5.2 Ethylene Regualtion of Transcription Factors............. 159
   8.5.3 Gene Expression during Softening...................... 160
   8.5.4 Genes Involved in Changes in Pigments and Volatiles....... 162
   8.5.5 Other Ripening-Related Genes Isolated by
          Differential Screening 162
8.6 Roles of Other Hormones and Metabolites during Ripening 163
   8.6.1 Auxin... 164
   8.6.2 Jasmonate and Jasmonic Acid 164
   8.6.3 Abscisic Acid 165
   8.6.4 Other Metabolites.................................... 166
8.7 Biotechnological Usage of Ethylene Biology.................... 167
   8.7.1 Transgenic Fruits with Altered Ethylene Production 167
   8.7.2 Transgenic Fruits with Altered Ethylene Perception 170
   8.7.3 Transgenic Fruits with Altered Fruit Softening............. 171
   8.7.4 Ripening-Related Promoters........................... 174
8.8 Conclusions 176
   References 176

9. Ethylene Involvement in Photosynthesis and Growth
Nafees A. Khan... 185
9.1 Introduction... 185
9.2 Plant Growth Regulators and Photosynthetic Responses 185
9.3 Ethylene in Photosynthesis and Growth of Mustard
   (*Brassica juncea*) 187
   9.3.1 Ethylene in Mustard Cultivars Differing In
          Photosynthetic Capacity 190

	9.3.1.1	*Effects of Ethylene Modulators*	191
	9.3.1.2	*Effects of ACC Synthase Activity Modulators*	193
9.4	Ethylene in Photosynthesis and Growth of Defoliated Plants		194
	References		197

Index 203

1 Interaction of Ethylene and Other Compounds with the Ethylene Receptor: Agonists and Antagonists

EDWARD C. SISLER[1], VARVARA P. GRICHKO[1], MARGRETHE SEREK[2]

1.1 Introduction

Although ethylene has long been recognized as a plant hormone, it is only recently that the ethylene receptor has been subjected to detailed study. Most reviews on ethylene signal transduction do not discuss much about ethylene interaction with the ethylene receptor except to mention that ethylene does bind to the receptor. This review will concentrate on the interaction of ethylene, ethylene agonists, and antagonists with the receptor. It is important that we identify the factors that determine compound binding and activity whether the compound is an ethylene agonist or an ethylene antagonist. It is important that findings from past work be noted in concordance with new-found results that contribute to our knowledge of the many compounds known to bind to the receptor. In recent years, the number and type of compounds that interact with the receptor has been expanded considerably. Some of these compounds appear to be useful both for basic research and for practical purposes. Many more may be discovered. It is the intent here to present some of what is known about both ethylene antagonists and agonists that have been found with the hope that the information will help lead to other compounds.

1.2 Ethylene and Agonists

1.2.1 Discovery of Ethylene Action and Some Important Lessons from the Past

Ethylene is one of the five original basic plant hormones. Many of the responses caused by ethylene were observed before it was known that it was the cause of the response (Abeles et al. 1992). In 1901, Neljubov reported that

---

[1] Department of Molecular and Structural Biochemistry, North Carolina State University, Campus Box 7622, Raleigh, NC 27695, USA

[2] Department of Agricultural Sciences, Crop Science, Royal Veterinary and Agricultural University, Thorvaldsensvej 40, 1871 Frederiksberg C, Denmark

ethylene caused a triple response in etiolated pea seedlings: i.e., epicotyl thickening, growth retardation, and horizontal growth of the epicotyl. It was soon recognized that ethylene was not alone in causing a triple response in plants. Soon it was known that propylene, acetylene, and carbon monoxide were ethylene agonists also giving a triple response in pea. In 1967, Burg and Burg identified several other alkenes and alkene-related compounds that were active. Isocyanides were added to the list of ethylene agonists 10 years later (Sisler 1977). These were important clues as to how ethylene may act. Now a great number of plant responses have been shown to be regulated by ethylene. Ethylene, which is produced by almost all plants, mediates a wide range of different plant responses and developmental steps. Ethylene plays an active role in seed germination, tissue differentiation, formation of root and shoot primordia, root elongation, lateral bud development, flowering initiation, anthocyanin synthesis, flower opening and senescence, pollination, fruit degreening and ripening, the production of volatile organic compounds responsible for aroma formation in fruits, leaf and fruit abscission, the response of plants to both biotic and abiotic stress, and plant-microbial interactions that are important for plant's growth and survival (Abeles et al. 1992; Grichko and Glick 2001a). Agricultural and horticultural loss is high due to ethylene-accelerated post-harvest ripening and deterioration of perishable commodities.

1.2.2 Molecular Requirements

In 1967, Burg and Burg published a paper on the molecular requirements for ethylene action in plants. Applying techniques used in enzyme kinetics, they compared a number of active compounds for their ability to give an ethylene agonistic response in peas. Using a gas chromatographic technique, they also compared the ability of the same compounds to bind to silver ions. They reported the binding of the compounds to silver ions to be in the same order as their ability to inhibit pea seedling growth. Burg and Burg (1967) then proposed that there was a metal in the supposed ethylene receptor. This was an important step toward understanding the way ethylene acts to bring about a response in plants. That ethylene binds to certain metals was not new. It had been known since 1827 that ethylene formed a complex with platinum and there was much chemical literature available on metal complexes of ethylene and other olefins, but the report by Burg and Burg (1967) was the first report extending this concept to plant responses. Consequently, there were several early attempts and suggestions to explain the mechanism of ethylene activity. Did ethylene act by being oxidized? Did ethylene act by producing some essential component as in an enzymatic reaction, or did ethylene serve to turn on a signal transduction pathway? Experimental evidence has favored a signal transduction pathway and this has been the focus of much recent work.

Some early work focused on the putative metal involved in ethylene action. Based on some deficiency experiments, Burg and Burg (1967) found that only zinc deficiency seemed to alter ethylene sensitivity in plants. For ethylene oxidation, copper seemed more likely as the metal. The reversible binding of ethylene to Cu(I) was well known (Cotton and Wilkinson 1980) and it seemed a likely prospect for being the metal involved (Sisler 1976, 1977). To support the proposed role of monovalent copper in the ethylene binding in plants, complexes of Cu(I) with imidazole-like ligands were synthesized (Thompson et al. 1983; Thompson and Whitney 1984; Thompson and Swiatek 1985). The complexes were the rather stable Cu(I) adducts with ethylene and its agonists and exhibited either a trigonal-planar geometry or a distorted tetrahedral structure. In a membrane environment, ligands bound to a metal ion may considerably alter its properties and the properties of its complex with ethylene, and there is a possibility that other metals might be involved in ethylene binding *in situ*. Rodriguez et al. (1999) did include other metals in an *in vitro* study where ethylene receptor gene *ETR1* from *Arabidopsis* was cloned in yeast. Only Cu(II) and Ag(I) significantly increased ethylene binding. Supplying ions such as Fe(II), Co(II), Ni(II), or Zn(II) did not increase ethylene binding. In the 2-D model of an ethylene receptor, which was developed based on these experiments, the transmembrane, hydrophobic ethylene-binding domain contained one Cu(I) ion per protein dimer, and coordinating amino acids were thought to be Cys65 and His69 (Rodriguez et al. 1999). The ethylene receptor has been suggested to contain either one or two Cu(I) ions per dimer (Hirayama et al. 1999; Pirrung 1999; Klee 2002; Taiz and Zeiger 2002; Weiler 2003). The stoichiometry gives little clue as to the structure. The coordination number of Cu(I) ions can be anywhere from two to six, and it is possible that Cu(I) forms a tetrahedral complex with both Cys65 and His69 (Pirrung 1999). A sulfur-ligated Cu(I)-ethylene complex exhibits very weak metal-ligand bonding interactions (Hirsh et al. 2001), and it is also possible that each cysteine residue is not a coordinating ligand. Cysteine residues instead may form disulfide bond *in situ*, and histidine residues and water may serve as ligands. Ethylene is likely to displace a weak ligand, and water is one of the most suitable candidates for this role. Displacement of water by ethylene followed by expelling of water molecule(s) from the hydrophobic domain is likely to result in the formation of a stable complex. An experiment in which a specific metal is shown to function *in situ* in ethylene perception has not yet been reported and is needed, but much recent evidence has favored copper as the metal involved in the receptor. Hirayama et al. (1999) restored antagonistic activity to an *Arabidopsis ran1* mutant, which gave an agonistic response with *trans*-cyclooctene (TCO), by either cloning a Cu(I) transporter into it or by supplying Cu(II) ions. This essentially confirms that copper can function in the receptor. The fact that *ran1* loss-of-function mutants were responsive to both ethylene and *trans*-cyclooctene is rather fascinating. Because 1-methylcyclopropene (1-MCP), a potent ethylene antagonist, also appeared to function normally, a metal must have been present in the receptor. Was that metal copper? *trans*-Cyclooctene was not included in the *in vitro* study of Rodriguez et al. (1999), and it is not known if the ethylene receptor associated with a different metal binds alkenes other than ethylene. Unusual behavior of the *ran1* mutant might be a result of either alteration of ethylene receptor conformation, decrease in ligand specificity, or the stability of receptors (Hirayama et al. 1999; Woeste and Kieber 2000). It can also be a result of irreversible disruption of altered ethylene receptors by *trans*-cyclooctene, enhanced sensitivity or insertion of different metal into some ethylene receptors under the conditions of a severe Cu(I) ion deficiency.

Some data suggest that binding of ethylene to the receptor may result in a structural rearrangement of the receptor, which can serve as an initial event in a signal transduction pathway. The role of histidine kinase activity of the ethylene receptor subfamily I is proven to be rather complex (Wang et al. 2003; Qu and Schaller 2004). It was shown that the ethylene receptor directly interacts with the downstream negative regulator CTR1 (Clark et al. 1998), and kinase activity of ETR1 is not required for its interaction with CTR1 (Gao et al. 2003).

There are still many questions about how ethylene acts. The exact structure of the ethylene-binding domain of the ethylene-receptor family is still unknown. The 3-D structure may be determined soon following a high-level expression of ETR1 in *E. coli* (Voet-van-Vormizeele and Groth 2003) and this may greatly facilitate the process of selection of the best candidates from the pool of synthetic compounds and phytochemicals and make it easier to predict anti-ethylene potency of their derivatives.

1.2.3 Ethylene Binding

Another important step in understanding the action of ethylene was the development of methods of measuring ethylene binding in plants. Using ^{14}C-ethylene, the rate of ethylene binding and the rate of ethylene release could be measured in plant tissue (Jerie et al. 1979; Sisler 1979). Using this technique, it could be shown that in vegetative tissue, there appeared to be a major component that bound and released ethylene rapidly. The time-radiolabeled ethylene remained bound to the major component varied in different plants. In most plants, the $t_{1/2}$ was about 10 min. However, in tomato leaflets it was only 2 min. The rapid component correlates well with the data of Warner and Leopold (1971) for response times by pea plants to ethylene. The value for K_d as determined by a Scatchard plot correlated well with the value for K_m as determined by a Lineweaver-Burk plot (Sisler 1979). There usually also was a small amount that was released with a much longer half-time. In some seeds, there were large amounts of ethylene, which remained bound for long periods of time. Because there is no known function for ethylene in these seeds, this probably represents binding to a storage component.

In measuring ethylene action, pea plants responded to ethylene in just 10 min and recovered with $t_{1/2}$ of about 18 min after its withdrawal (Warner and Leopold 1971). In *Arabidopsis* hypocotyls of etiolated seedlings, there were two phases of growth inhibition by ethylene, a rapid phase followed by a prolonged slower phase. Full recovery occurs about 90 min after ethylene removal (Binder et al. 2004a). The recovery time was significantly smaller than the time of ethylene dissociation from ETR1 receptors expressed in yeast (Schaller and Bleecker 1995; Binder et al. 2004a). The inhibition appears to be a complex process (Binder et al. 2004b). In ethylene binding studies, the shortest value of $t_{1/2}$ for ^{14}C ethylene diffusion from the binding site measured *in vivo* was 2 min (Sisler 1982).

In vitro, the short-lived component is absent; in extracts of mung bean sprouts, $t_{1/2}$ of 1 h and $t_{1/2}$ of 50 h were measured (Sisler 1990). In a cell-free system from cotyledons of *Phaseolus vulgaris*, $t_{1/2}$ was about 10.5 h (Bengochea et al. 1980). In yeast expressing ETR1 at a level of about 4.0×10^{-8} M, $t_{1/2}$ was 12.5 h (Schaller and Bleecker 1995). Based on K_d and $t_{1/2}$ (Sisler 1991), one can estimate that the rate constant of ethylene binding to the receptor is about 5×10^7 M^{-1} s^{-1} for the short-lived component and 2×10^5 M^{-1} s^{-1} for the long-lived component, indicating that rate of ethylene binding is likely to be determined by the rate of its interaction with the active center of the receptor.

1.3 Ethylene Antagonists

1.3.1 Chemical Adjuvants Counteracting Ethylene

Ethylene responses in plants can be prevented to some extent by a number of chemical adjuvants. High concentrations of sucrose, carbon dioxide, and cycloheximide delay senescence in flowers (Dilley and Carpenter 1975). Carbon dioxide is used in controlled atmosphere storage of fruits and vegetables and it has been known for many years that it is a natural inhibitor of ethylene responses. Early studies of the carbon dioxide effect suggested that it competes with ethylene in ethylene action (Burg and Burg 1967); however, direct measurement with ^{14}C-labeled ethylene did not indicate that carbon dioxide competes with ethylene for the receptor sites (Sisler 1979). Recently it was shown that carbon dioxide acts by suppressing ethylene biosynthesis (John 1997). Indoleacetic acid can prevent ethylene action under some circumstances. Application of indoleacetic or 2,4-dichlorophenoxyacetic acid to plant tissue will retard some ethylene-induced processes, but there is no evidence that they act by preventing ethylene binding, and their action also seems to be indirect (Sisler et al. 1985).

Responses to ethylene are controlled by either lowering its biosynthesis or limiting its action. A number of inhibitors of ethylene biosynthesis have been developed. Ethylene biosynthesis in plants can also be minimized by expression of a microbial ACC deaminase gene or genetic suppression of the key enzymes of the Yang cycle (Klee et al. 1991; Theologis et al. 1992). For example, ACC deaminase transgenic tomato plants that are resistant to flooding stress may be constructed by using root-specific promoters, which are also anaerobically inducible (Grichko and Glick 2001b). The practical disadvantages of genetic approaches are the necessity for development of transgenic lines of each species, which is almost impossible, and a high public concern associated with the issues of transgenic food. A non-invasive and universal way of controlling ethylene responses in plants is emerging. Indeed, plant growth can be affected in a variety of ways by plant growth-promoting bacteria expressing the ACC deaminase gene (Grichko and Glick 2001a). Because these approaches affect ethylene biosynthesis and do not protect plants from exogenous ethylene, in recent years much effort has been focused on the control of ethylene action that starts with the binding of ethylene to the receptor (Sisler 1979; Schaller and Bleecker 1995). Ag(I) ion (Beyer 1976) especially silver thiosulfate is a very effective inhibitor of ethylene action. Ag(I) ion interacts with the receptor and binds ethylene in this state (Rodriguez et al. 1999) but fails to induce response *in situ*. The Ag(I) ion is thought to occupy the binding site of the receptor (Rodriguez et al. 1999) or it might affect it in some other way. The silver ion reacts with sulfur-containing compounds and is known to deactivate enzymes by reacting with sulfhydryl groups. Silver, being a heavy metal, has been banned from use to counteract ethylene in some countries and this limits its use.

1.3.2 Ethylene Agonists That Require Continuous Exposure to Give a Response

All existing ethylene antagonists except for silver thiosulfate and cyclopropenes require continuous exposure. The fact that some alkenes inhibit ethylene responses was discovered by Sisler and Pian in 1973. 2,5-Norbornadiene (2,5-NBD) was known to form one of the most stable silver complexes. Out of curiosity, 2,5-NBD was tested on tobacco leaves, flowers, and seeds to see if it would elicit an ethylene response. It did not appear to induce an ethylene-like response but instead did seem to overcome the effect of ethylene. Several other cyclic alkenes were then tested and were effective inhibitors of ethylene action. 2,5-NBD was the best antagonist found among the cyclic alkenes. However, it required continuous exposure and had a pungent and obnoxious odor. Despite these limitations, for many years it served as an important experimental tool, and it initiated the search for ways to control the ethylene receptor. One of the more important results found with the cycloalkenes was that the level of activity appeared to depend on the ring strain (Sisler and Yang 1984). The more strained the alkene, the better it was as an antagonist. 2,5-NBD continued to be the best alkene antagonist until *trans*-cyclooctene was discovered (Sisler et al. 1990). TCO is not much more highly strained than 2,5-norbornadiene and concentration-wise it was nearly 100 times as effective. TCO also has a very pungent and obnoxious odor and must be prepared by synthesis. It has had only limited usage. These compounds remain bound much longer than ethylene, diffusing from the binding site with a $t_{1/2}$ of 3–6 h (Sisler et al. 1990). Many ethylene responses require more than 6 h of exposure to ethylene for induction of an observable response, and the results of a single exposure would not be sufficient to be noted. This is probably the reason continuous exposure is required. Cyclic alkenes that are potent ethylene antagonists are listed in Table 1.1. Cyclopentadiene had been found to be about as effective as 2,5-NBD as an inhibitor of ethylene responses. Cyclobutene also proved to be a compound requiring continuous exposure.

Table 1.1. Inhibition of ethylene action in plants by competitive antagonists

Compound name	Structure	Plant	K_i (µL L^{-1} gas)
Diazocyclopentadiene		Carnation	0.12
trans-Cyclooctene		Banana	0.78
4-Penten-1-ol		Banana	110
cis-Cyclooctene		Banana	512
2,5-Norbornadiene		Pea Banana	170 55
Cyclopentadiene		Banana	140

(*Continued*)

Table 1.1. Inhibition of ethylene action in plants by competitive antagonists—(cont'd)

Compound name	Structure	Plant	K_i (µL L^{-1} gas)
Allylbenzene		Banana	189
4-Phenyl-1-butene		Banana	206
Norbornene		Pea	360
1,3-Cyclohexadiene		Pea	488
2-Vinylnaphthalene		Banana	490
1,3-Cycloheptadiene		Pea	870
2-Allylphenol		Banana	995
Cyclopentene		Pea	1,100
1,4-Cyclohexadiene		Pea	4,650
Cyclohexene		Pea	6,060

Table 1.1. Inhibition of ethylene action in plants by competitive antagonists—*(cont'd)*

Compound name	Structure	Plant	K_i (µL L^{-1} gas)
Cyclohexane		Pea	Inactive
Benzene		Pea	Inactive

*After Sisler (1991), Grichko et al. (2003). K_i value is the amount of the compound required to double K_m. The lower the K_i value is, the more effective the inhibitor

1.3.3 Naturally Occurring Ethylene Antagonists

Surprisingly, the ethylene-binding domain seems to be rather easily accessible by relatively large naturally occurring molecules that contain ten or more carbons (Grichko et al. 2003). Many naturally occurring terpenes compete with ethylene for the receptor. All of the terpenes tested had some antagonistic activity. There was considerable variation in the potency found, but some were in the range where they might be important in nature (Table 1.2). In particular, when an oxygen atom was near the double bond the compound was usually more active than one not having an oxygen atom. This was true whether the oxygen-containing group was an aldehyde, hydroxyl, or keto group. The effect might be due to a hydrophilic interaction or it could be because the oxygen group withdrew electrons from the double bond causing the molecule to bind more tightly. All tested terpenes required continuous exposure for activity and some of these were comparable with 2,5-NBD in activity. It is not known if antagonistic properties of terpenes are important in nature in controlling ethylene responses, but it is possible. Many natural compounds exhibit allelopathic inhibition of seed germination and growth of competing plants (Fischer et al. 1994), and basic research on naturally occurring ethylene antagonists may reveal presently unknown mechanisms underlying such complex phenomena as allelopathy, growth inhibition of neighboring species, and regulation of plant growth and development. Possibly ethylene action could be involved in these phenomena.

1.3.4 Photoactivated Compounds

Research on photoactivated antagonists was undertaken by Sisler et al. (1993) to label and identify the ethylene receptor. Because UV photolysis of diazo compounds frequently results in the formation of carbenes that rapidly react with other compounds including proteins, diazocyclopentadiene (DACP) was synthesized and found to bind to the ethylene receptor under continuous exposure. However, UV light did not activate DACP, and severe damage was done to the

Table 1.2. Ethylene antagonists of plant origin

Name	Structure	K_i (µL L)*
Monoterpenes		
Linalool		101
(+) Carvone		103
Perillaldehyde		167
Carveol		337
α-Pinene		360
α-Terpinene		400

Table 1.2. Ethylene antagonists of plant origin—(cont'd)

Name	Structure	K_i (μL L)*
γ-Terpinene		400
Limonene		588
Perillalcohol		693
Myrcene		1,333
β-Pinene		3,580
Isoprene		22,000

(Continued)

Table 1.2. Ethylene antagonists of plant origin—(cont'd)

Name	Structure	K_i (μL L)*
Products of phenylpropanoid and other pathways		
Eugenole		101
Estragole		103
trans-Cinnamaldehyde		165
cis-2-Hexene-1-ol		175
trans-2-Hexene-1-ol		195
Cinnamyl alcohol		821

*After Grichko et al. (2003). Data are given for inhibiting ethylene responses in banana fruit. Concentrations are given for the compound as a gas. K_i values are the amount required to double K_m The lower the K_i value the greater the antagonistic effect

plants. It was found though that DACP was much more active after exposure to fluorescent light than before, and it appeared to be an excellent inhibitor of ethylene action (Table 1.1) (Sisler and Blankenship 1993). Attempts to radiolabel DACP and identify the active photolytic products by GC analysis failed. The active product(s) remain unknown. The explosiveness of DACP limited its use.

1.4 Ethylene Agonists That Require a Single Exposure

1.4.1 Ring Strain and Cyclopropene

Similar to enzyme inhibitors, full antagonists that possess zero efficacy often exhibit very high affinity toward the receptor (Levitzki 1984). Apparent K_d of the native ethylene-receptor complex is about 10^{-10} M (Bengochea et al. 1980;

Sisler 1991), and the tendency of other alkenes to form complexes with Cu(I) and Ag(I) that are more stable than complexes with ethylene is strongly correlated with their anti-ethylene properties. Strain appears to be an important factor that determines the potency of antagonists, but alone it is not sufficient to cause a compound to be an ethylene antagonist. For example, cyclopropane has a high ring strain value but does not seem to interact with the ethylene receptor because it has no double bond. It is neither an ethylene agonist nor an antagonist (Sisler, unpublished). Methylenecyclopropane has high ring strain but the double bond is located outside the ring, and it is an ethylene agonist. 2,5-NBD and TCO are very potent ethylene antagonists and very strained compounds (Muhs and Weiss 1962; Sisler 1991). Continuing with the concept of ring strain being the important property of all potent ethylene antagonists, more strained compounds that could be the ethylene antagonists were synthesized. Cyclopropene (CP) was found to be very effective as an inhibitor of ethylene responses (Table 1.3) but has only undergone limited testing. CP boils at −35 °C and is very unstable. It can polymerize with explosiveness if warmed too rapidly (Schipperijn and Smael 1973). It must be stored at very low temperatures but can be stored at room temperature for extended periods of time as a gas in an inert atmosphere in the absence of solvent. It rapidly decays in solvents at room temperature.

1.4.2 1-Methylcyclopropene

1-Methylcyclopropene (1-MCP) is more stable than CP. However, it is also an unstable compound in the liquid state or in solution except at low temperatures. No data on boiling point has been found for it. Based on the boiling points of similar compounds, it should be near 0 °C. In the gaseous state at low concentrations under an inert atmosphere, 1-MCP is stable at room temperature for months. 1-MCP is available commercially. A 24-h exposure to 0.5–0.7 nL L^{-1} 1-MCP is sufficient to protect carnation flowers, banana fruits, and tomato fruits (Table 1.3). A single exposure protects banana fruits against ethylene for approximately 12 days at 23 °C. Some tissues, such as etiolated pea plants, require approximately 40 nL L^{-1} for protection. The large concentration difference (about 80 fold) may be due to the different rate of synthesis of ethylene receptors but there is no experimental evidence for this (Sisler et al. 1996a, 1996b). There are no data on the natural occurrence of 1-MCP.

1.4.3 3-Methylcyclopropene

3-Methylcyclopropene (3-MCP) (bp −4 °C) is an effective inhibitor of ethylene action. It also appears to be somewhat less stable than 1-MCP. Although 3-MCP has the same empirical formula as 1-MCP, it is biologically less effective. Depending on the plant material, the effective concentration of 3-methylcyclopropene required was reported to be from five to ten times

Table 1.3. Minimal concentration of cyclopropenes and time of imposed insensitivity on banana fruits[*]

Cyclopropene name	Short name	Structure	Concentration (nL L^{-1} gas)	Time (days)
Cyclopropene	CP		0.7	12
1-Methylcyclopropene	1-MCP		0.7	12
3-Methylcyclopropene	3-MCP		2	12
3,3-Dimethylcyclopropene	3,3-DMCP		500	7
3-Methyl-3-vinylcyclopropene	3,3-MVCP		120	5
1,3,3-Trimethylcyclopropene	1,3,3-TCP		20,000	12
3-Methyl-3-ethynylcyclopropene	3,3-MECP		240	5
1,3-Dimethylcyclopropene	1,3-DCP		250	12
1,2-Dimethylcyclopropene	1,2-DCP		3,000	3
1-Ethylcyclopropene	1-ECP		4	12
1-Propylcyclopropene	1-PCP		6	12
1-Butylcyclopropene	1-BCP		3	12
1-Pentylcyclopropene	1-PentCP		0.5	14
1-Hexylcyclopropene	1-HCP		0.4	20
1-Heptylcyclopropene	1-HeptCP		0.4	21
1-Octylcyclopropene	1-OCP		0.3	25
1-Nonylcyclopropene	1-NCP		0.4	35
1-Decylcyclopropene	1-DCP		0.3	36

[*]After Sisler et al. (2001, 2003). Minimal concentration of the cyclopropene gas is the concentration that is necessary to give protection of chlorophyll degradation for extended periods of time. Time of protection against ethylene is number of days during which banana fruits remained insensitive to ethylene after exposure to saturating amount of cyclopropene. Exposure to the cyclopropene was 24 h

higher than that required for 1-MCP (Sisler et al. 1999). It is not known if this is due to the presence of the methyl group in the 3 position or the lack of a methyl group in the 1 position or both. It does show the effect of structure on biological activity. Protection time on bananas at 23 °C is 12 days.

1.4.4 3,3-Dimethylcyclopropene

3,3-Dimethylcyclopropene (3,3-DMCP) (bp 14.5 °C) is a very stable compound. It can be kept at 100 °C for many hours without decomposition (Closs 1966). 3,3-DMCP is biologically 1,000 times less active than 1-MCP (Table 1.3). Protection time on bananas is 7 days. 3,3-DMCP may be of practical value if short time protection of agricultural commodities against ethylene is desired. 3,3-DMCP has only undergone limited testing.

1.4.5 Other 1-Alkyl Cyclopropenes

The compounds listed in Table 1.3 that are substituted in the 1 position are usually protected longer than those substituted in the 3 position. It was decided to explore the further extending of the substitution in the 1 position. Burg and Burg (1967) had shown that the ratio of activities for ethylene, propylene, and 1-butene was 1:130:27,000 and it was supposed that the same sort of ratio would apply to inhibitors since Burg and Burg (1967) interpreted their results as the ethylene binding site was restricted and only small molecules could bind to it. A series of compounds were prepared with the side chain in the 1 position being extended from 1–10 carbons. As the side chain was extended, the antagonistic potency declined, but beyond four carbons, potency increased again above that of CP and 1-MCP, and continued to be high until the ten carbon chain was attached. In addition to higher activity, the protection time on bananas increased from 12 to 36 days for a single exposure (Feng et al. 2004). 1-Methyl-, 1-ethyl-, 1-propyl-, 1-butyl-, 1-pentyl-, 1-hexyl-, 1-heptyl-, 1-octyl-, 1-nonyl, and 1-decylcyclopropene are all very potent antagonists (Sisler et al. 2003). Also, 0.7 nL L^{-1} 1-MCP protected banana fruits against ethylene after a 24-h exposure, but many 1-substituted cyclopropenes require even lower concentrations than 1-MCP. 1-MCP, 1-ethyl-, 1-propyl-, and 1-butylcyclopropene protected banana for 12 days, 1-pentylcyclopropene protected bananas for 14 days, 1-hexylcyclopropene for 20 days, 1-heptylcyclopropene for 21 days, 1-octylcyclopropene for 25 days, 1-nonylcyclopropene for 35 days, and 1-decylcyclopropene for 36 days (Sisler et al. 2003). The binding site did not appear to be restricted and the larger compounds seemed to be bound stronger than the small ones. It appeared that there could be more than one kind of interaction involved. These compounds were all highly strained, but hydrophobic interaction could be also important.

1.4.6 Other Cyclopropenes

Cyclopropenes are ethylene antagonists that are all effective upon a single exposure. Because inhibition by cyclopropenes is irreversible (Sisler et al. 1996a, 1996b), it is possible that it occurs via the covalent attachment of the cyclopropene moiety to the susceptible, active site of the receptor, but it is thought that a coordination bond is their mode of attachment because the range of inactivation time ranges from 3 to 36 days, depending on the cyclopropene, and this follows a pattern that is more likely due to a hydrophobic effect. It is not known if cyclopropenes stay intact or undergo ring-opening *in situ*. Both the formation of complexes with ring retention and ring cleavage followed by the formation of copper-carbon bond is possible (Visser et al. 1973; Halton and Banwell 1987). Cyclopropenes are known to react with thiols forming alkylthiocyclopropane adducts, and cyclopropene fatty acids and cyclopropenols irreversibly inhibit iron-containing fatty acyl CoA desaturases by the reaction of the cyclopropene ring with a cysteine residue at the active site of the enzyme (Reiser and Raju 1964; Quintana et al. 1998; Triola et al. 2001; Rodriguez et al. 2004). Is it possible that the similar mechanism takes place in a case of ethylene receptor and reaction of the alkylcyclopropene with Cys65 does irreversibly upset the binding center in the receptor? Although the exact mechanism of their action is not known, cyclopropenes seem to form the most stable adducts with the receptor and, as a consequence, these compounds are the most effective ethylene antagonists with the long-lasting effect (Sisler et al. 1996a, 1996b; Sisler 2002).

Results with 1-MCP and other light cyclopropenes pointed to the need to know more about the effect of different substituents on activity of cyclopropene. Are the large differences in concentration requirements and binding time due to structural considerations or are they due to electronic effects, or both? To try to answer these questions, a number of compounds were prepared and tested. As different structures were used, some definite trends were noted. Relative to hydrogen, methyl, vinyl, and ethynyl groups are electron donating. Two substituents on the cyclopropene ring increased concentration requirements and decreased protection time. When two substituents were present in the 3 position, protection time was reduced unless the 1 position was also substituted. Substitution in the 1 position usually increased binding time. These results suggest that both steric effects and electronic effects are important, and usually the presence of substituents seem to weaken the compounds.

1.4.7 Is the Binding Site Restricted?

Since results on large 1-cyclopropenes indicated that the binding site in ethylene receptor was not restricted, it was reasonable to expand early experiments performed with light alkenes (Burg and Burg 1967) to dodecene. In pea, data obtained for ethylene, propylene and 1-butene were in good agreement with those obtained previously by Burg and Burg (1967). Increase in molecular size decreases agonistic properties: ethylene gives half-maximum response at $0.1\ \mu L\ L^{-1}$, propylene is effective at $10\ \mu L\ L^{-1}$ and 1-butene gives half-maximum response at $27\ mL\ L^{-1}$. Larger alkenes did not exhibit agonistic properties, but 1-butene, 1-pentene, 1-hexene, 1-octene, 1-decene, and

1-dodecene were found to be competitive inhibitors of ethylene action (Sisler 2004). As the size of hydrocarbon chain increased, the potency of antagonist increased. 1-Butene is a pivotal compound exhibiting both the agonistic properties and antagonistic properties. These results are rather unexpected, and it is likely that more than one mechanism is involved in this phenomenon. However, it clearly demonstrates that the ethylene-binding site is not very restricted.

1.4.8 Hydrophobic Interaction

Strain appears to be the major factor in the binding of cyclic alkenes to ethylene receptors in plants, but with alkenes, hydrophobic interaction is probably the major factor. With 1-substituted cyclopropenes both ring strain and a hydrophobic interaction appear to be involved (Sisler et al. 2003). Because the structure of the ethylene-binding center in the ethylene receptor and the mechanism of interaction of both ethylene and its antagonists with the receptor have not been yet established, it is not always possible to predict the exact effect of the molecular structure on antagonistic properties. Notwithstanding, the hydrophobic molecule that contains a long, saturated chain on a double bond is likely to be excellent membrane-targeting agent. The presence of the hydrophobic unit in the molecule of an antagonist, which is able to interact with the hydrophobic domain of the receptor, should increase its binding affinity. Indeed, it was found that hydrophobic substituents substantially enhance anti-ethylene properties of ethylene antagonists and increase the period of imposed insensitivity to ethylene (Sisler et al. 2003). After withdrawal from the atmosphere, large hydrophobic compounds incorporated into the lipid membrane should stay there for a longer time than gaseous 1-MCP or light antagonists and such compounds are expected to continue blocking the newly synthesized receptors and postpone tissue recovery. In a similar way, sterculate that is larger than malvalate has a higher preference for incorporation into membranes than malvalate (Pawlowski et al. 1985). On the other hand, hydrophilic substituents may allow discrimination between different plant tissues.

The modeling and selection of potential ethylene antagonists has always been based on the coordination chemistry of d^{10} metals and their interaction with alkenes, i.e., both Cu(I)-alkene and Ag(I)-alkene complexes. There are many ways one can modify the potency of ethylene antagonists, which depends on many factors, including molecule size, shape, strain, functional groups, degree of saturation, and hydrophobicity. The position, number, and geometry of double bonds and near functional groups are all important with regard to the antagonist potency (Sisler et al. 2001). In addition, electron-withdrawing and electron-donating substituents may affect biological activity of antagonists in different ways, allowing time-resolved control of ripening. Thus, electron-withdrawing substituents reduce efficacy of agonists;

the relative effect of ethylene decreases along the series ethylene, vinyl chloride, and carbon monoxide. The increase in molecular size enhances antagonistic properties and positively correlates with the protection time (Sisler 2002; Grichko et al. 2003).

1.5 How Does an Agonist Start a Signal?

Ethylene, carbon monoxide, isocyanides, allene, acetylene, propylene and other alkenes that induce an agonistic response are π-acceptors. These ligands are capable of accepting some of the electron density from metal ions into their empty π or π^* orbitals. Sisler (1977) proposed a ligand substitution governed by the *trans* effect as a model for the ethylene mode of action. In the trans effect, a ligand tends to facilitate substitution in the position trans to itself (Cotton and Wilkinson 1980). Exactly what happens *in vivo* is unknown. The first step in signal transduction must be the binding of ethylene to the putative metal in the receptor. Ethylene, like its congeners, will form a complex and back-accept electron density into its vacant orbital. This will change the distribution of the charge on the metal and might start a rearrangement of ligands associated with the metal. Somehow, this signal is transmitted along the signal transduction pathway in such a way that it causes a response. All of the compounds that give an ethylene response probably do it in essentially the same way (Fig. 1.1). Once the rearrangement takes place, what happens to the receptor? This is not known. Compounds that compete with ethylene for the receptor, but do not themselves induce a response, probably initially act in the same way (Fig. 1.2). For example, in some *Arabidopsis* mutants, 2,5-NBD and TCO can act as partial agonists. One of the big questions is: Why can ethylene and its agonists turn on the ethylene response signal but ethylene antagonists such as 2,5-NBD, *trans*-cyclooctene, and cyclopropenes cannot? How do the compounds differ? Inhibitors bind and likely withdraw electrons starting a ligand rearrangement but they remain on the receptor longer suggesting that it must become free to be active. These antagonists are strained compounds and likely bind tighter and, thus, occupy

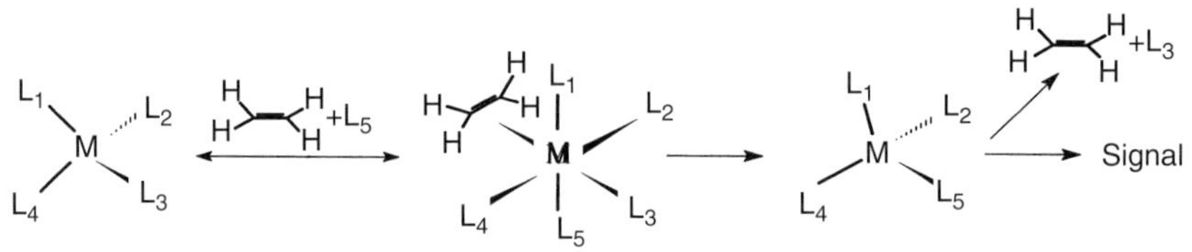

Fig. 1.1. Possible mechanism of ethylene perception via the rearrangement of ligands in the metal active center. The model only accounts for the first step in the signal transduction pathway proposed elsewhere (Clark et al. 1998; Rodriguez et al. 1999)

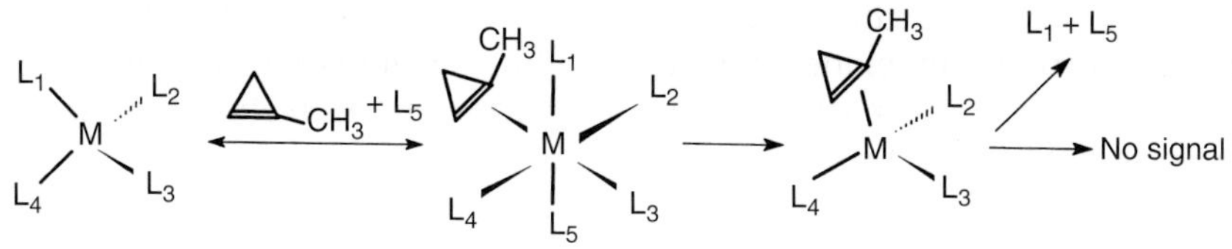

Fig. 1.2. When 1-MCP binds no signal is produced because ligand rearrangement cannot be completed

the receptor longer than unstrained ones. Is it the time of receptor occupancy that is critical? Among 1-alkenes, ethylene, propylene, and 1-butene are agonists, and 1-pentene, 1-hexene, 1-decene and 1-dodecene are antagonists. These alkenes differ only by the molecular size, and the antagonistic effect increases as it increases. Strain, if any, should be the same. A hydrophobic effect and increase in the time when receptor is occupied would be likely. It is known that the *in vivo* $t_{1/2}$ of ethylene-receptor complex is about 10 min and the $t_{1/2}$ of 2,5-NBD and TCO dissociation from the receptor is about 3 and 6 h, respectively. Cyclopropenes remain bound significantly longer. When the antagonist binds to the metal, it appears that it cannot complete the rearrangement and locks the receptor in an inactive state.

1.6 Selected Papers on Significance of Ethylene Antagonists

Since the discovery of inhibitors of ethylene action the ethylene antagonists have become a powerful tool that helps researchers to make significant advances in many fields of plant biology. Both new gene discovery and elucidation of mechanisms of gene expression, and the characterization of ethylene-binding sites and identification of new components of ethylene signal transduction pathway were tremendously accelerated by the introduction of ethylene antagonists and especially by the discovery of 1-MCP.

1.6.1 Application in Molecular Biology

In plants, ethylene-regulated genes can comprise as much as 7% of a genome (van Zhong and Burns 2003), and ethylene antagonists allow many advanced approaches to accelerated gene discovery and protein characterization. For example, a new approach to the identification of novel components of ethylene perception and signal transduction pathway was introduced that is based on the development of mutants showing agonist-like responses to the potent ethylene antagonists (Hirayama et al. 1999). A long-term exposure of *Arabidopsis* plants to *trans*-cyclooctene resulted in the development of *responsive-to-agonist1* (*ran1*) mutant exhibiting ethylene phenotype in response to treatment with both *trans*-cyclooctene and ethylene. It was found that *RAN1* encodes a copper-transporting P-type ATPase that is similar to yeast Ccc2p and human Wilson and Menkes proteins. The RAN1 is thought to deliver copper to secure the synthesis of fully functional ethylene receptors. The isolation of *enhanced ethylene-response* (*eer*) mutants in *Arabidopsis* has revealed that 2,5-NBD also can induce ethylene-like responses in plants (Larsen and Chang 2001). Similar to RAN1, EER1 is likely to act upstream of ethylene receptors. At high levels, 2,5-NBD was reported to act as ethylene agonist in deep-water rice (Bleecker et al. 1987). Because 2,5-NBD induces ethylene-like responses either in plants with enhanced sensitivity or when applied at high concentrations, it can be formally classified as a partial agonist with the low efficacy. All these findings show that ethylene antagonists are particularly useful for developing mutants with altered perception and their study.

In mature green, breaker, orange, and red ripe tomato fruits, 1-MCP treatment decreased the mRNA abundance of phytoene synthase 1, expansin 1, and ACC oxidase 1, three ripening-related tomato genes (Hoeberichts et al. 2002). Therefore the ripening process can be inhibited both on a physiological and a molecular level, even at very advanced stages. In carnation flowers, transcription of three members of ACC synthase gene family (*DCACS1, DCACS2, and DCACS3*) and *ACO1* (ACC oxidase 1 gene) is decreased by 1-MCP (Jones 2003), confirming that ethylene is the primary regulator of *ACS* and *ACO* gene expression in some flower parts. In agreement with the previous work, the Northern analysis of gene expression of ACC synthase and the ACC oxidase showed that expression of both genes declined in DACP-treated tomato fruit and then recovered. It was concluded that the effects of DACP on ethylene biosynthesis are on expression of ACC synthase and ACC oxidase genes, and/or regulation of ACC oxidase activity (Tian et al. 1997). In banana fruit, an interesting observation was the effect of 1-MCP on ACC content and *in vitro* ACC oxidase activity (Pathak et al. 2003). Though 1-MCP-treated banana fruits did not show any increase in ethylene production, 1-MCP did not inhibit completely both ACC accumulation and ACC oxidase. No expression of ACC synthase was observed in 1-MCP-treated banana fruits, whereas a basal level of ACC oxidase transcript was detected throughout. It was suggested that ripening of banana is different from that of other climacteric fruits, and that ethylene biosynthesis may be rather complex during ripening. In some cases, the effect of 1-MCP on ACC synthase activity can be further enhanced by high levels of CO_2 (Lu and Toivonen 2003). Itai et al. (2003) studied accumulation of mRNAs for β-D-xylosidase and α-L-arabinofuranosidase during tomato fruit development. Using 1-MCP it was shown that expression of the LeARF1 gene is negatively regulated by ethylene and expression of the β-D-xylosidase gene is independent of ethylene. In experiments with branchlets of broccoli, Gong and Mattheis (2003) showed that 1-MCP reduced activity of chlorophyllase and peroxidase and did not affect the activity of lipoxygenase and proposed that chlorophyll in broccoli can be degraded via the peroxidase-hydrogen peroxide system. In apricots,

1-MCP altered α-D-galactosidase, β-D-galactosidase, α-D-mannosidase, and α-D-glucosidase activities, inhibited ethylene biosynthesis, delayed softening and did not affect fruit color (Botondi et al. 2003). Ethylene stimulated activities of pectin methyl esterase, polygalacturonase, pectate lyase and cellulase in banana fruits while 1-MCP suppressed the ethylene effect (Lohani et al. 2004). In pear fruit, the mRNA accumulation of polygalacturonase genes was in parallel with the pattern of fruit softening in 1-MCP treatment. However, the expression pattern of pear endo-1,4-β-D-glucanase genes was not affected, suggesting that ethylene is required for polygalacturonase expression even in the late ripening stage but not for endo-1,4-β-D-glucanase genes (Hiwasa et al. 2003).

Ethylene antagonists are intensively used to study properties of different proteins and establish the role that both these proteins and other regulatory components may play in plant metabolism. Thus, 1-MCP was successfully used to biochemically characterize the ethylene receptors ETR1 and ERS1 from *Arabidopsis* following the heterologous expression in yeast (Hall et al. 2000). In the study of ethylene biosynthesis, the treatment of fruits with 1-MCP helped to clarify the role of carbon dioxide. Kinetic parameters derived from the models pointed to the production of ethylene from ACC by ACC oxidase as a possible action site for carbon dioxide inhibition (de Wild et al. 2003). Involvement of ethylene in the response of rice to elevated CO_2 levels was confirmed in experiments with 1-MCP (Seneweera et al. 2003). 1-MCP had only a small effect on alternative oxidase in green pepper, a non-climacteric fruit, which was selected to investigate if low temperature and ethylene are involved in regulation of mitochondrial respiration and alternative oxidase (AOX). Exogenous ethylene stimulated mitochondrial respiration, transcription, and translation of existing AOX, but did not induce a new isozyme of AOX (Tian et al. 2004). In Steinite et al. (2004), inhibitor studies conducted with 1-MCP and aminooxyacetic acid indicated that hydrogen peroxide generated through NADPH oxidase and superoxide dismutase is necessary for regurgitant-induced increase of ethylene production and oxidative enzyme activities in bean leaves. 1-MCP has affected antioxidant enzymes activity and antioxidant content of the peel of apples that may play an important role in preventing the superficial scald in apples (Shaham et al. 2003). In Friedman et al. (2003) 1-MCP helped to demonstrate that octanoic acid does not augment ethylene response in *Arabidopsis*, though it might affect ethylene action in flower abscission of the ethylene-overproducer mutant. To address potential signaling interactions between herbivore-induced jasmonic acid and ethylene, Schmelz et al. (2003) pretreated corn plants with 1-MCP. This decreased production of ethylene and volatile emission following beet armyworm caterpillars herbivory but did not affect the accumulation of jasmonic acid. The results support a role for jasmonic acid in the regulation of insect-induced volatile emission and suggest that ethylene regulates the magnitude of volatile emission during herbivory. In *Petunia inflata*, pollination triggers two phases of ethylene production by the pistil, the first of which peaks 3 h after pollination. To investigate the physiological significance of the first phase, pollinated flowers were treated with 2,5-NBD and 1-MCP. Both antagonists caused an inhibition of pollen tube growth during the first 6 h of pollination, indicating that pistil tissues are the primary target of the pollination-induced ethylene (Holden et al. 2003).

1.6.2 Application for Control of Plant Growth and Development

Ethylene antagonists are intensively used to control plant growth and development. Early papers on the application of diazocyclopentadiene, 1-MCP, other cyclopropenes, and other blockers of ethylene action in plants were extensively reviewed elsewhere (Sisler and Serek 1997, 1999, 2003; Blankenship and Dole 2003). The compounds preventing ethylene responses have been developed into a means for protecting plants against ethylene and extending the shelf life of some commodities. 1-MCP is now commercially available under the names EthylBloc and SmartFresh and is currently used on flowers, fruits and vegetables with great success. During ripening, 1-MCP-treated fruits attain quality similar to that of controls (Argenta et al. 2003). Moreover, treatment with 1-MCP can potentially maintain some intrinsic functional food quality parameters (i.e., antioxidant levels) in stored fruit (MacLean et al. 2003). In ethylene sensitive flowers, among other responses, it prevents senescence and abscission of plant organs; in fruits and vegetables, it slows down the ripening process. Other cyclopropene inhibitors are now being developed for a range of application (Sisler and Serek 2003). Recently it was shown that 1-MCP stimulates organogenesis of kiwi explants (Arigita et al. 2003) and, thus, can be used as a regulator of *in vitro* organogenesis.

1-MCP found numerous applications in amelioration of stress responses in plants. In Yokotani et al. (2004) responses of wild-type, *rin* and *nor* tomato fruits to both ethylene and wounding were studied to identify the events controlled by each mutation. Treatment with 1-MCP, preceding wounding, inhibited expression of *E4* but did not affect the expression of other genes in the lines tested. It was concluded that a wounding signal that controls *LE-ACS2*, *LE-ACS6*, and *LE-ACO1* is independent of *rin* and *nor* mutations and ethylene signaling (Yokotani et al. 2004). Another example includes treatment of broccoli with 1-MCP alone or in combination with 0.2 µL L^{-1} ozone that maintained the quality and extended the shelf life of broccoli florets. In addition, the 1-MCP treatment maintained a higher chlorophyll fluorescence ratio during the prolonged storage and reduced production of dimethyl trisulfide, which contributes to off-odor in broccoli florets (Forney et al. 2003). Grimmig et al. (2003) found that 1-MCP blocked the ethylene-induced grapevine resveratrol synthase gene expression in transgenic tobacco leaves, but did not affect ozone-induced expression. It was proposed that ozone-induced gene expression occurs via different signaling mechanisms suggesting an additional ethylene independent signaling pathway for ozone-induced expression of genes involved in phytoalexin biosynthesis. 1-MCP prevented cold-water stress in tomato plants that usually results in elevated ethylene production and development of dwarfing in seedlings (Huang and Lin 2003). 1-MCP reduced mesocarp discoloration, decay development, and polyphenol oxidase activity in ethylene-treated avocado fruits that were cold stored for prolonged period of time (Pesis et al. 2002).

With some precaution, 1-MCP can be beneficial for inhibiting the degreening process in citrus fruits (Porat et al. 1999). The beneficial effects of 1-MCP were observed in both pre-climacteric and climacteric apple fruits (Fan et al. 1999). 1-MCP inhibited apple ripening and loss of firmness and titratable acidity when fruits were held at 0 °C up to 6 months, and when the fruits were held at 20–24 °C, for up to 2 months. Storage and shelf life were extended for all cultivars tested, and ethylene production and respiration were reduced substantially by the treatment. Interestingly, 1-MCP-treated apple fruits had soluble solids equal to or greater than those in non-treated fruits (Fan et al. 1999). Custard apples that have a very short storage life at room temperature also showed higher firmness than the control fruits following treatment with 1-MCP (Benassi et al. 2003). A single application of 1-MCP can efficiently retard post-harvest senescence of parsley leaves (Ella et al. 2003) and slow down softening in persimmon cultivars (Harima et al. 2003). There are some certain advantages in combining 1-MCP with other technologies used in food preservation. For example, avocado fruits treated with both 1-MCP and wax had better retention of green peel color and fruit firmness, and delayed climacteric ethylene evolution and respiration rates compared with other treatments (Jeong et al. 2003). 1-MCP treatment was effective in reducing ethylene production, respiration, and loss of firmness and color of slices when applied to whole apples directly after harvest while total sugar and acidity levels were not affected by the treatment (Perera et al. 2003). Also, 1-MCP treatment helps maintain the quality of minimally processed pineapple fruits at least partially by reducing the hydrolysis of endogenous ascorbic acid (Budu and Joyce 2003). Application of 1-MCP improves the maintenance of the green color of skin in lime fruits (Kluge et al. 2003) and increases storability and shelf life in climacteric and nonclimacteric plums (Martinez-Romero et al. 2003). Ripe green banana fruits treated with 1-MCP significantly delayed the peaks of respiration rate and ethylene production but did not reduce the peak height. Reductions in firmness, acidity, and starch content of banana fruits were remarkably delayed by 1-MCP treatment. 1-MCP treatment also delayed the increase of total soluble sugar and soluble pectin contents while the soluble solids content in treated fruit remained almost unchanged during the storage (Jiang et al. 2004a). When 1-MCP was applied to 'Bartlett' pears prior to storage at 10 °C, the synergistic interaction of low temperature and 1-MCP resulted in an extended post-harvest life after transfer to room temperature, with concomitant delay in ethylene production, respiration rate, and color development, and retention of firmness (Trinchero et al. 2004). Treatment of apple trees with 1-MCP was shown to decrease fruit drop (Sato et al. 2004). When Santa Rosa plums were treated, 1-MCP ethylene and CO_2 production were strongly inhibited, higher values of firmness were observed; the treatment delayed color evolution, reduced acidity loss, and inhibited ethanol and acetaldehyde production (Salvador et al. 2003). 1-MCP did not affect weight loss or sugar loss. However, 1-MCP did not have any effect on shelf life of blueberry (DeLong et al. 2003) and had only a small effect on total storage life of strawberries (Bower et al. 2003). 1-MCP inhibited ethylene-induced ripening of avocado fruit at very low concentrations (Feng et al. 2000).

1-MCP and other cyclopropenes robustly inhibit exogenous ethylene effects in flowers such as bud and flower drop, leaf abscission, and flower senescence, and are used to control senescence in flowers (Serek et al. 1994, 1995). Treatments with 1-MCP significantly prolonged vase life of *Cattleya alliances* (Yamane et al. 2004). Lilies may benefit from pretreatment with 1-MCP when cut stems contain buds that are marginally small for opening and when the stems will be cold stored before marketing (Han and Miller 2003). Sweet pea flowers have a very short post-harvest life and are sensitive to exogenous ethylene. Flowers treated with either 200 nL L^{-1} 1-octylcyclopropene or 200 nL L^{-1} 1-MCP for 6 h were protected against ethylene, and their display life was prolonged up to almost 7 days (Kebenei et al. 2003b). Both compounds were shown to be excellent blockers of ethylene responses in sweet pea flowers. 1-MCP inhibited petal abscission in *Pelargonium peltatum* (Cameron and Reid 2001) though its effect was thought to depend on both shipping and storage temperature and application frequency.

1.6.3 Commercial Use

The commercial products that are based on 1-MCP, the first patented nontoxic ethylene action inhibitor, were introduced a few years ago by the USA companies FloraLife Inc. and AgroFresh Inc., a Rohm and Haas Company, under the names EthylBloc and SmartFresh. EthylBloc is aimed mainly at ornamental crops, while SmartFresh is for edible commodities. Both products have been developed as powders with 1-MCP complexed with γ-cyclodextrin, which, when mixed with water or buffer solution, releases the gas. For commercial use, plant material has to be treated in enclosed areas such as tightly built greenhouses, rooms, coolers, truck trailers, shipping boxes/containers, etc. The treatment areas should be gas-tight as much as possible for preventing gas leakage, which will reduce the effectiveness of 1-MCP. Tapes or other products or procedures are recommended to prevent leakage as well as for the establishment of an internal air circulation system during the treatment (without bringing outside air in). Once the plant material is treated it does not need re-treatment, however re-treatment is not harmful and can even be beneficial. Some species of ornamental plants would likely benefit from additional treatment, especially with flowers in different stages of development on the same plant, which are continuously developing new ethylene binding sites.

In many scientific studies performed under laboratory conditions it has been documented that only very small amounts of 1-MCP, in the range of nL L^{-1}, are necessary for preventing ethylene responses (Fig. 1.3). However, the recommended concentration for EthylBloc and SmartFresh is in the range of μl L^{-1}, 1,000 fold higher, which probably takes into consideration a high possibility for leakage of 1-MCP. The recommended temperature during the treatment is based on several scientific reports as well as on practical trials: for ornamental crops not lower than 13 °C, for edible products 1.6–24 °C. Treatments at lower temperature require increased concentration of 1-MCP. Both EthylBloc and SmartFresh were qualified for review in the USA. The Environmental Protection Agency's Biopesticide division refers to them as

Fig. 1.3. Examples of use of 1-MCP for preventing of ethylene effects **a** ripening (chlorophyll degradation) in tomato fruits, **b** sleepiness of kalanchoe flowers, **c** leaf yellowing in geranium, **d** bud abscission in Christmas cactus, **e** growth of etiolated pea seedlings (plants were treated with 20 nL/L 1-MCP for 6 h followed by treatment with the indicated amount of ethylene) **f** flower senescence in carnation

reduced-risk products due to their extremely low usage rate and highly favorable safety profile. They leave no detectable residues on fruits, which is another positive attribute of using this technology. Concentration of residues, if any, is so low that it is considered below detection limit. As such, the EPA granted an exemption from tolerance for the products. Both products can eliminate the need for a residual crop protection treatment used today to control a common storage disorder, which is an additional benefit.

EthylBloc was registered for ornamentals in 1999 for use in the USA and later also in Canada. In the Netherlands, the product was recently registered for tulip bulbs. Registration of SmartFresh for selected edible products like apples, avocados, persimmons, tomatoes, papaya, plums has already been granted in the USA, Mexico, Chile, Argentina, South Africa, New Zealand, Israel, Brazil, the Netherlands, and the UK. The registration in several other countries, especially EU countries, for a range of Horticultural products is in progress. Additional information can be obtained at: http://www.rohmhaas.com/ethylbloc

1.6.4 Limitations of Compounds

Both agriculture and horticulture are excited over the potential impact of ethylene antagonists that block the receptors since they protect plants from both endogenous and exogenous ethylene, and can be applied to any ethylene-sensitive plant material. One of the best ethylene antagonists so far is silver thiosulfate, but its toxicity limits its use (Beyer 1979). There are some substantial drawbacks associated with the practical application of 1-MCP. 1-MCP is absorbed differently by different commodities and that should be taken into consideration when prescribing commercial application of the antagonist (Dauny et al. 2003). The 1-MCP effect on fruit ripening is temperature-dependent and is enhanced by low temperature storage and reduced by the high temperature storage (Jiang et al. 2004a). In some cases, special care must be taken to avoid chilling injuries and decay development in fruits (Porat et al. 1999). However, 1-MCP is most effective at delaying ripening of mature-green tomatoes when they are stored near a temperature range of 12.5–15 °C (Mostofi et al. 2003). Unfortunately, in some cases 1-MCP also accelerates ethylene production in some plants. The effect of 1-MCP in inducing massive ethylene biosynthesis is rapid and reaches a maximum within just a few hours (Ella et al. 2003). The finding that 1-MCP can promote ethylene biosynthesis in parsley, freshly harvested pears and other commodities (Ella et al. 2003; de Wild et al. 2003) warrants special care in its application. The high amount of ethylene emitted by 1-MCP-treated plants may lead to post-harvest damage to other mixed-load commodities that are sensitive to ethylene but are not protected against it. However, ethylene production by 1-MCP-treated kiwi fruit remained low, and the fruits did not show an climacteric ethylene production even after 32 days of storage at room temperature (Boquete et al. 2004). The same phenomenon was observed in pears (Kubo et al. 2003; Jiang et al. 2004b) when 1-MCP treatments suppressed ethylene and carbon dioxide production significantly and slowed fruit softening. In *Cattleya alliances* flowers, 1-MCP suppressed ACC oxidase activity and ethylene production during the first 2 days after the treatment and doubled the vase life; ethylene production did substantially increase after the fourth day (Yamane et al. 2004). There is some concern about the change in fruit aroma that may be induced by ethylene antagonists. Thus, biosynthesis of monoterpenes, esters, and aldehydes in the mango fruit is strongly dependent on ethylene production and action and is suppressed by 1-MCP (Lalel et al. 2003). The respiration rate, ethylene production, and fatty acids content of the fruit during ripening were also decreased in 1-MCP-treated fruits. In some cases, if fruits have started to ripen they are relatively insensitive to 1-MCP although treatment with 1-MCP reduces fruit sensitivity to handling damage, even after ripening (Ekman et al. 2004). In contrast, continuous application of 1-MCP to tomato fruits completely inhibited color development of breaker and half-ripe fruits and partially inhibited firmness loss that may make 1-MCP usage commercially feasible (Mir et al. 2004). 1-MCP may increase bitter rod and blue mold decay in apples, which can be prevented by using biocontrol agents such as *Metschnikowia pulcherrima* (Janisiewicz et al. 2003). Not all ethylene-sensitive flowers respond strongly to the treatment with 1-MCP. For instance, repeated 1-MCP treatments provided only a modest extension in longevity of attached daffodil flowers held in air and had no noticeable effect on the life of detached flowers held in water (Hunter et al. 2004).

It was found that the effect of 1-MCP on ethylene production significantly depends on the treatment conditions. For example, ethylene production in peaches was not affected by 1-MCP at 20 °C but was suppressed after application at 0 °C (Liguori et al. 2004). 1-Octylcyclopropene was efficient in preventing senescence in *Kalanchoë* flowers at 15 and 20 °C while lower temperatures reduced its potency (Kebenei et al. 2003a). 1-MCP appeared to have a limited effect on some stone fruit species and cultivars. The effect is concentration and storage duration-dependent (Argenta et al. 2003). In a similar study, 1-MCP was also found to be much less effective in delaying ripening of partially ripened bananas (Pelayo et al. 2003). However, 1-MCP treatment, after the initiation of pear fruit ripening, markedly suppressed the subsequent flesh softening and ethylene production (Hiwasa et al. 2003). In another study, banana fruits treated with 1-MCP and then held in an atmosphere enriched with oxygen followed by exposure to ethylene softened more rapidly than fruits held in air. Accelerated softening of fruits exposed to high levels of oxygen may suggest that oxygen stimulates synthesis of new ethylene binding sites (Jiang and Joyce 2003). 1-MCP may affect the volatiles profile of fruits, especially in apricots, reducing the synthesis of lactones and promoting the rise of terpenols (Botondi et al. 2003). The effect of 1-MCP is limited by its instability and, in some cases, by its short-term residual activity. To approach this problem, a simple 1-MCP sustained release device that prolongs 1-MCP exposure was recently introduced by Macnish et al. (2004).

1.6.5 Future Needs

Ripening of fruits treated with 1-MCP and its analogues is inhibited for a finite period of time after which the fruits ripen normally (Feng et al. 2004). However, 1-MCP has a number of significant drawbacks that may limit its usage. 1-MCP is a very unstable gas with a long-term effect that cannot be terminated when desired. The duration of the imposed insensitivity to ethylene upon the treatment of plant material with 1-MCP varies substantially in different commodities. 1-MCP is nearly insoluble in water. Its commercial use requires enclosed areas and it was found rather ineffective at low temperatures. This makes the treatment of agricultural commodities with 1-MCP difficult inside the low-temperature storage facilities as anticipated by the industry. In addition, some naturally occurring cyclopropenes are known to inhibit fatty acid desaturation and gluconeogenesis in animals and possess carcinogenic and neurochemical activity (Pawlowski et al. 1985; Salaun and Baird 1995). Thus there is a stable demand in the development of cyclopropene inhibitors of ethylene action that can retard fruit ripening and flower senescence for the different periods of time and have desirable physical and chemical properties such as, for example, particular solubility in water, volatility, stability, hydrophobicity, ring strain energy, and etc. (Kebenei et al. 2003a). 1-Hexylcyclopropene and 1-octylcyclopropene are found to be potent ethylene inhibitors. Pre-treatment of *Kalanchoë* flowers with these compounds delayed inrolling and extended the display life of flowers. 1-Octylcyclopropene prolonged the display life of flowers to almost 10 days and was significantly better than 1-MCP applied at the same concentration (Kebenei et al. 2003a). To meet the standards of the $10 billion organic food industry, the development of natural ethylene receptor inhibitors is highly desirable (Grichko et al. 2003). A natural ethylene action inhibitor should be a non-toxic compound of plant origin that can be applied at a rather low concentration. It would be practical to develop an ethylene receptor blocker that could counteract ethylene action for a desired period of time to achieve temporary suspension of fruit ripening and flower senescence.

1.7 Concluding Remarks

For many years, only a few small molecules were known to interact with the ethylene receptor to give ethylene responses. Now many compounds are known to act at the receptor level, some of these giving ethylene responses and some acting to counteract ethylene. Some of these compounds are synthetic chemicals and some are naturally occurring compounds. Our knowledge of the factors involved in ethylene receptor antagonist action is only partial, and there are some factors that have not been examined. There are many types of compounds that have not been tested. The effects of functional groups on activity of the compounds have not been examined in detail. For future developments, it is important that these and possibly other factors be better understood. Different ways of applying and evaluating chemicals need to be developed. It is impossible to predict what future work will bring but the prospects seem good that more will come.

Acknowledgements: Support for work in the authors' laboratory from the United States-Israel Binational Agricultural Research and Development Fund (BARD US 3493–03C) is gratefully acknowledged.

References

Abeles FB, Morgan PW, Saltveit ME Jr (1992) Ethylene in plant biology, 2nd edn. Academic, New York
Argenta LC, Krammes JG, Megguer CA, Amarante CVT, Mattheis J (2003) Ripening and quality of 'Laetitia' plums following harvest and cold storage as affected by inhibition of ethylene action. Pesqui Agropecu Bras 38:1139–1148
Arigita L, Tames RS, Gonzalez A (2003) 1-Methylcyclopropene and ethylene as regulators of *in vitro* organogenesis in kiwi explants. Plant Growth Regul 40:59–64
Beyer EM Jr (1976) A potent inhibitor of ethylene action in plants. Plant Physiol 58:268–271
Beyer EM Jr (1979) Effect of silver ion, carbon dioxide, and oxygen on ethylene action and metabolism. Plant Physiol 63:169–173
Benassi G, Correa GASF, Kluge RA, Jacomino AP (2003) Shelf life of custard apple treated with 1-methylciclopropene: an antagonist to the ethylene action. Braz Arch Biol Tec 46:115–119
Bengochea T, Dodds JH, Evans DE, Jerie PH, Niepel B, Shaari AR, Hall MA (1980) Studies on ethylene binding by cell-free preparations from cotyledons of *Phaseolus vulgaris* L. Planta 148:397–406
Binder BM, O'Malley RC, Wang W, Moore JM, Parks BM, Spalding EP, Bleecker AB (2004a) *Arabidopsis* seedling growth response and recovery to ethylene: a kinetic analysis. Plant Physiol 136:2913–2920
Binder BM, Mortimore LA, Stepanova AN, Ecker JR, Bleecker AB (2004b) Short-term growth responses to ethylene in *Arabidopsis* seedlings are EIN3/EIL1 independent. Plant Physiol 136:2921–2930
Blankenship SM, Dole JM (2003) 1-methylcyclopropene: a review. Postharvest Biol Technol 28:1–25
Blankenship SM, Sisler EC (1993) Use of diazocyclopentadiene to block ethylene action in fruits and flowers. Acta Hort 343:215–216
Bleecker AB, Rose-John S, Kende H (1987) An evaluation of 2,5-norbornadiene as a reversible inhibitor of ethylene action in deep-water rice. Plant Physiol 84:395–398
Boquete EJ, Trinchero GD, Fraschina AA, Vilella F, Sozzli GO (2004) Ripening of 'Hayward' kiwifruit treated with 1-methylcyclopropene after cold storage. Postharvest Biol Technol 32:57–65
Botondi R, DeSantis D, Bellincontro A, Vizovitis K, Mencarelli F (2003) Influence of ethylene inhibition by 1-methylcyclopropene on apricot quality, volatile production, and glycosidase activity of low- and high-aroma varieties of apricots. J Agr Food Chem 51:1189–1200
Bower JH, Blasi WV, Mitcham EJ (2003) Effects of ethylene and 1-MCP on the quality and storage life of strawberries. Postharvest Biol Technol 28:417–423
Budu AS, Joyce DC (2003) Effect of 1-methylcyclopropene on the quality of minimally processed pineapple fruit. Aust J Exp Agr 43:177–184
Burg SP, Burg EA (1967) Molecular requirement for the biological activity of ethylene. Plant Physiol 42:144–152
Cameron AC, Reid MS (2001) 1-MCP blocks ethylene-induced petal abscission of *Pelargonium peltatum* but the effect is transient. Postharvest Biol Technol 22:169–177

Clark KL, Larsen PB, Wang X, Chang C (1998) Association of *Arabidopsis* CTR1 Raf-like kinase with the ETR1 and ERS ethylene receptors. Proc Natl Acad Sci USA 95:5401-5406

Closs GL (1966) Cyclopropenes. In: Advances in alicyclic chemistry. Academic, New York, vol. 1, pp 53-127

Cotton FA, Wilkinson G (1980) Advanced inorganic chemistry. Wiley, Chichester, pp 763-766

Dauny PT, Joyce DC, Gamby C (2003) 1-Methylcyclopropene influx and efflux in 'Cox' apple and 'Hass' avocado fruit. Postharvest Biol Technol 29:101-105

DeLong JM, Prange RK, Bishop C, Harrison PA, Ryan DAJ (2003) The influence of 1-MCP on shelf-life quality of highbush blueberry. Hort Sci 38:417-418

De Wild HPJ, Otma EC, Peppelenbos HW (2003) Carbon dioxide action on ethylene biosynthesis of preclimacteric and climacteric pear fruit. J Exp Bot 387:1537-1544

Dilley DR, Carpenter WJ (1975) The role of chemical adjuvants and ethylene synthesis on cut flower longevity. Acta Hort (ISHS) 41:117-132

Ekman JH, Clayton M, Biasi WV, Mitcham EJ (2004) Interactions between 1-MCP concentration, treatment interval and storage time for 'Bartlett' pears. Postharvest Biol Technol 31:127-136

Ella L, Zion A, Nehemia A, Amnon L (2003) Effect of the ethylene action inhibitor 1-methylcyclopropene on parsley leaf senescence and ethylene biosynthesis. Postharvest Biol Technol 30:67-74

Fan XT, Blankenship SM, Mattheis JP (1999) 1-methylcyclopropene inhibits apple ripening. J Am Soc Hort Sci 124:690-695

Feng XQ, Apelbaum A, Sisler EC, Goren R (2000) Control of ethylene responses in avocado fruit with 1-methylcyclopropene. Postharvest Biol Technol 20:143-150

Feng XQ, Apelbaum A, Sisler EC, Goren R (2004) Control of ethylene activity in various plant systems by structural analogues of 1-methylcyclopropene. Plant Growth Regul 42:29-38

Fischer NH, Williamson GB, Weidenhamer JD, Richardson DR (1994) In search of allelopathy in the Florida scrub: the role of terpenoids. J Chem Ecol 20:1355-1379

Forney CF, Song J, Fan LH, Hildebrand PD, Jordan MA (2003) Ozone and 1-methylcyclopropene alter the postharvest quality of broccoli. J Am Soc Hort Sci 128:403-408

Friedman H, Meir S, Philosoph-Hadas S, Halevy AH (2003) Effect of octanoic acid on ethylene-mediated processes in *Arabidopsis*. Plant Growth Regul 40:239-247

Gao ZY, Chen YF, Randlett MD, Zhao XC, Findell JL, Kieber JJ, Schaller GE (2003) Localization of the Raf-like kinase CTR1 to the endoplasmic reticulum of *Arabidopsis* through participation in ethylene receptor signaling complexes. J Biol Chem 278:34725-34732

Gong YP, Mattheis JP (2003) Effect of ethylene and 1-methylcyclopropene on chlorophyll catabolism of broccoli florets. Plant Growth Regul 40:33-38

Grichko VP, Glick BR (2001a) Amelioration of flooding stress by ACC deaminase-containing plant growth-promoting bacteria. Plant Physiol Biochem 39:11-17

Grichko VP, Glick BR (2001b) Flooding tolerance of transgenic tomato plants expressing the bacterial enzyme ACC deaminase controlled by the *35S*, *rolD* or *PRB*-1b promoter. Plant Physiol Biochem 39:19-25

Grichko VP, Sisler EC, Serek M (2003) Anti-ethylene properties of monoterpenes and some other naturally occurring compounds in plants. SAAS Bull Biochem Biotech 16:20-27

Grimmig B, Gonzalez-Perez MN, Leubner-Metzger G, Vogeli-Lange R, Meins F, Hain R, Penuelas J, Heidenreich B, Langebartels C, Ernst D, Sandermann H (2003) Ozone-induced gene expression occurs via ethylene-dependent and -independent signaling. Plant Mol Biol 51:599-607

Halton B, Banwell MG (1987) Cyclopropenes. In: Rappoport Z (ed) The chemistry of the cyclopropyl group. Wiley, Chichester, pp 1223-1339

Hall AE, Findell JL, Schaller GE, Sisler EC, Bleecker AB (2000) Ethylene perception by the ERS1 protein in *Arabidopsis*. Plant Physiol 123:1449-1457

Han SS, Miller JA (2003) Role of ethylene in postharvest quality of cut Oriental lily 'Stargazer'. Plant Growth Regul 40:213-222

Harima S, Nakano R, Yamauchi S, Kitano Y, Yamamoto Y, Inaba A, Kubo Y (2003). Extending shelf life of astringent persimmon (*Diospyros kaki* Thunb.) fruit by 1-MCP. Postharvest Biol Technol 29:319-324

Hirayama T, Kieber JJ, Hirayama N, Kogan M, Guzman P, Nourizadeh S, Alonso JM, Dailey WP, Dancis A, Ecker JR (1999) Responsive-to-antagonist1, a Menkes/Wilson disease-related copper transporter, is required for ethylene signaling in *Arabidopsis*. Cell 97:383–393

Hirsh J, George SD, Solomon EI, Hedman B, Hodgson KO, Burstyn JN (2001) Raman and extended X-ray absorption fine structure characterization of a sulfur-ligated Cu(I) ethylene complex: modeling the proposed ethylene binding site of *Arabidopsis thaliana* ETRI. Inorg Chem 40:2439–2441

Hiwasa K, Kinugasa Y, Amano S, Hashimoto A, Nakano R, Inaba A, Kubo Y (2003) Ethylene is required for both the initiation and progression of softening in pear (*Pyrus communis* L.) fruit. J Exp Bot 54:771–779

Hoeberichts FA, Van der Plas LHW, Woltering EJ (2002) Ethylene perception is required for the expression of tomato ripening-related genes and associated physiological changes even at advanced stages of ripening. Postharvest Biol Technol 26:125–133

Holden MJ, Marty JA, Singh-Cundy A (2003) Pollination-induced ethylene promotes the early phase of pollen tube growth in *Petunia inflata*. J Plant Physiol 160:261–269

Huang JY, Lin CH (2003) Cold water treatment promotes ethylene production and dwarfing in tomato seedlings. Plant Physiol Bioch 41:283–288

Hunter DA, Yi MF, Xu XJ, Reid MS (2004) Role of ethylene in perianth senescence of daffodil (*Narcissus pseudonarcissus* L. 'Dutch Master'). Postharvest Biol Technol 32:269–280

Itai A, Ishihara K, Bewley JD (2003) Characterization of expression, and cloning, of β-D-xylosidase and α-L-arabinofuranosidase in developing and ripening tomato (*Lycopersicon esculentum* Mill.) fruit. J Exp Bot 54:2615–2622

Janisiewicz WJ, Leverentz B, Conway WS, Saftner A, Reed AN, Camp MJ (2003) Control of bitter rot and blue mold of apples by integrating heat and antagonist treatments on 1-MCP treated fruit stored under controlled atmosphere conditions. Postharvest Biol Technol 29:129–143

Jeong H, Huber DJ, Sargent SA (2003) Delay of avocado (*Persea americana*) fruit ripening by 1-methylcyclopropene and wax treatments. Postharvest Biol Technol 28:247–257

Jerie PH, Shaari AR, Hall MA (1979) The compartalization of ethylene in developing cotyledons of *Phaseolus vulgaris* L. Planta 144:503–507

Jiang W, Zhang M, He J, Zhou L (2004a) Regulation of 1-MCP-treated banana fruit quality by exogenous ethylene and temperature. Food Sci Technol Int 10:15–20

Jiang WB, Sheng Q, Jiang YM, Zhou XJ (2004b) Effects of 1-methylcyclopropene and gibberellic acid on ripening of Chinese jujube (*Zizyphus jujuba* M) in relation to quality. J Sci Food Agr 84:31–35

Jiang YM, Joyce DC (2003) Softening response of 1-methylcyclopropene-treated banana fruit to high oxygen atmospheres. Plant Growth Regul 41:225–229

John P (1997) Ethylene biosynthesis: the role of 1-aminocyclopropane-1-carboxylate (ACC) oxidase, and its possible evolutionary origin. Physiol Plant 100:583–592

Jones ML (2003) Ethylene biosynthetic genes are differentially regulated by ethylene and ACC in carnation styles. Plant Growth Regul 40:129–138

Kebenei Z, Sisler EC, Winkelmann T, Serek M (2003a) Efficacy of new inhibitors of ethylene perception in improvement of display life of kalanchoë (*Kalanchoë blossfeldiana* Poelln.) flowers. Postharvest Biol Technol 30:169–176

Kebenei Z, Sisler EC, Winkelmann T, Serek M (2003b) Effect of 1-octylcyclopropene and 1-methylcyclopropene on vase life of sweet pea (*Lathyrus odoratus* L.) flowers. J Hort Sci Biotech 78:433–436

Klee HJ (2002) Control of ethylene-mediated processes in tomato at the level of receptors. J Exp Bot 53:2057–2063

Klee HJ, Hayford MB, Kaetzmer KA, Barry GF, Kishore GM (1991) Control of ethylene synthesis by expression of a bacterial enzyme in transgenic tomato plants. Plant Cell 3:1187–1193

Kluge RA, Jomori MLL, Jacomino AP, Vitti MCD, Padula M (2003) Intermittent warming in 'Tahiti' lime treated with an ethylene inhibitor. Postharvest Biol Technol 29:195–203

Kubo Y, Hiwasa K, Owino WO, Nakano R, Inaba A (2003) Influence of time and concentration of 1-MCP application on the shelf life of pear 'La France' fruit. Hort Sci 38:1414–1416

Lalel HJD, Singh Z, Tan SC (2003) The role of ethylene in mango fruit aroma volatiles biosynthesis. J Hort Sci Biotech 78:485–496

Larsen PB, Chang C (2001) The Arabidopsis *eer1* mutant has enhanced ethylene responses in the hypocotyl and stem. Plant Physiol 125:1061–1073
Levitzki A (1984) Receptors: a quantitative approach. Benjamin/Cummings Publ, Menlo Park, California, pp 103–105
Liguori G, Weksler A, Zutahi Y, Lurie S, Kosto I (2004) Effect of 1-methylcyclopropene on ripening of melting flesh peaches and nectarines. Postharvest Biol Technol 31:263–268
Lohani S, Trivedi PK, Nath P (2004) Changes in activities of cell wall hydrolases during ethylene-induced ripening in banana: effect of 1-MCP, ABA and IAA. Postharvest Biol Technol 31:119–126
Lu CW, Toivonen PMA (2003) 1-Methylcyclopropene plus high CO_2 applied after storage reduces ethylene production and enhances shelf life of Gala apples. Can J Plant Sci 83:817–824
Macnish AJ, Joyce DC, Irving DE, Wearing AH (2004) A simple sustained release device for the ethylene binding inhibitor 1-methylcyclopropene. Postharvest Biol Technol 32:321–338
MacLean DD, Murr DP, DeEll JR (2003) A modified total oxyradical scavenging capacity assay for antioxidants in plant tissues. Postharvest Biol Technol 29:183–194
Martinez-Romero D, Dupille E, Guillen F, Valverde JM, Serrano M, Valero D (2003) 1-Methylcyclopropene increases storability and shelf life in climacteric and nonclimacteric plums. J Agr Food Chem 51:4680–4686
Mir N, Canoles M, Beaudry R, Baldwin E, Pal Mehla C (2004) Inhibiting tomato ripening with 1-methylcyclopropene. J Am Soc Hort Sci 129:112–120
Mostofi Y, Toivonen PMA, Lessani H, Babalar M, Lu CW (2003) Effects of 1-methylcyclopropene on ripening of greenhouse tomatoes at three storage temperatures. Postharvest Biol Technol 27:285–292
Muhs MA, Weiss FT (1962) Determination of equilibrium constants of silver-olefin complexes using gas chromatography. J Am Chem Soc 84:4697–4705
Neljubov DN (1901) Uber die horizontale Nutation der Stengel von Pisum sativum und einiger anderen Pflanzen. Beih Bot Centralbl 10:128–139
Pathak N, Asif MH, Dhawan P, Srivastava MK, Nath P (2003) Expression and activities of ethylene biosynthesis enzymes during ripening of banana fruits and effect of 1-MCP treatment. Plant Growth Regul 40:11–19
Pawlowski NE, Hendricks JD, Bailey ML, Nixon JE, Bailey GS (1985) Structural-bioactivity relationship for tumor promotion by cyclopropenes. J Agric Food Chem 33:767–777
Pelayo C, Vilas-Boas EVD, Benichou M, Kader AA (2003) Variability in responses of partially ripe bananas to 1-methylcyclopropene. Postharvest Biol Technol 28:75–85
Perera CO, Balchin L, Baldwin E, Stanley R, Tian M (2003) Effect of 1-methylcyclopropene on the quality of fresh-cut apple slices. J Food Sci 68:1910–1914
Pesis E, Ackerman M, Ben-Arie R, Feygenberg O, Feng XQ, Apelbaum A, Goren R, Prusky D (2002) Ethylene involvement in chilling injury symptoms of avocado during cold storage. Postharvest Biol Technol 24:171–181
Pirrung MC (1999) Histidine kinases and two-component signal transduction systems. Chem Biol 6:R167–R175
Porat R, Weiss B, Cohen L, Daus A, Goren R, Droby S (1999) Effects of ethylene and 1-methylcyclopropene on the postharvest qualities of 'Shamouti' oranges. Postharvest Biol Technol 15:155–163
Qu X, Schaller GE (2004) Requirement of the histidine kinase domain for signal transduction by the ethylene receptor ETR1. Plant Physiol 136:2961–2970
Quintana J, Barrot M, Fabrias G, Camps F (1998) A model study on the mechanism of inhibition of fatty acyl desaturases by cyclopropene fatty acids. Tetrahedron 54:10187–10198
Reiser R, Raju PK (1964) Inhibition of saturated fatty acid dehydrogenation by dietary fat containing sterculic + malvalic acids. Biochem Biophys Res Commun 17:8–13
Rodriguez FI, Esch JJ, Hall AE, Binder BM, Schaller GE, Bleecker AB (1999) A copper cofactor for the ethylene receptor ETR1 from *Arabidopsis*. Science 283:996–998
Rodriguez S, Camps F, Fabrias G (2004) Inhibition of the acyl-CoA desaturases involved in the biosynthesis of *Spodoptera littoralis* sex pheromone by analogs of 10,11-methylene-10-tetradecenoic acid. Insect Biochem Mol 34:283–289

Salaun J, Baird MS (1995) Biologically active cyclopropanes and cyclopropenes. Curr Med Chem 2:511–542
Salvador A, Cuquerella J, Martinez-Javega JM (2003) 1-MCP treatment prolongs postharvest life of 'Santa Rosa' plums. J Food Sci 68:1504–1510
Sato T, Kudo T, Akada T, Wakasa Y, Niizeki M, Harada T (2004) Allelotype of a ripening-specific 1-aminocyclopropane-1-carboxylate synthase gene defines the rate of fruit drop in apple. J Am Soc Hort Sci 129:32–36
Schaller GE, Bleecker AB (1995) Ethylene-binding sites generated in yeast expressing the *Arabidopsis ETR1* gene. Science 270:1809–1811
Schipperijn AJ, Smael P (1973) Chemistry of cyclopropene II. Formation and reactions of 1-potassio, 1-sodio and 1-lithiocyclopropene in liquid ammonia. Recueil 92:1121–1166
Schmelz EA, Alborn HT, Banchio E, Tumlinson JH (2003) Quantitative relationships between induced jasmonic acid levels and volatile emission in Zea mays during *Spodoptera exigua* herbivory. Planta 216:665–673
Seneweera S, Aben SK, Basra AS, Jones B, Conroy JP (2003) Involvement of ethylene in the morphological and developmental response of rice to elevated atmospheric CO_2 concentrations. Plant Growth Regul 39:143–153
Serek M, Sisler EC, Reid MS (1994) Novel gaseous ethylene-binding inhibitor prevents ethylene effects in potted flowering plants. J Am Soc Hort Sci 119:1230–1233
Serek M, Tamari G, Sisler EC, Borochov A (1995) Inhibition of ethylene-induced cellular senescence symptoms by 1-methylcyclopropene, a new inhibitor of ethylene action. Physiol Plant 94:229–232
Shaham Z, Lers A, Lurie S (2003) Effect of heat or 1-methylcyclopropene on antioxidative enzyme activities and antioxidants in apples in relation to superficial scald development. J Am Soc Hort Sci 128:761–766
Sisler EC (1976) Ethylene analogs: effect of some unsaturated sulfides (thioethers) on tobacco leaves. Tob Sci 20:6–10
Sisler EC (1977) Ethylene activity of some π-acceptor compounds. Tob Sci 21:43–45
Sisler EC (1979) Measurement of ethylene binding in plant tissue. Plant Physiol 64:538–542
Sisler EC (1982) Ethylene binding in normal, *rin* and *nor* mutant tomatoes. Plant Growth Regul 1:219–226
Sisler EC (1990) Ethylene-binding receptors: is there more than one? In: Pharis RP, Rood SB (eds) Plant growth substances 1988. Springer, Berlin Heidelberg New York, pp 192–200
Sisler EC (1991) Ethylene-binding components in plants. In: Matoo AK, Suttle JC (eds) The plant hormone ethylene. CRC Press, Boca Raton, pp 81–99
Sisler EC (2002) Methods of blocking ethylene response in plants using cyclopropene derivatives. US Patent 6,365,549, 2 April 2002
Sisler EC (2004) Interaction of compounds with the ethylene binding site. PGRSA Q 32:52
Sisler EC, Blankenship SM (1993) Diazocyclopentadiene (DACP), a light sensitive reagent for the ethylene receptor in plants. Plant Growth Regul 12:125–132
Sisler EC, Pian A (1973) Effect of ethylene and cyclic olefins on tobacco leaves. Tob Sci 17:68–72
Sisler EC, Serek M (1997) Inhibitors of ethylene responses in plants at the receptor level: recent developments. Physiol Plant 100:577–582
Sisler EC, Serek M (1999) Compounds controlling the ethylene receptor. Bot Bull Acad Sinica 40:1–7
Sisler EC, Serek M (2003) Compounds interacting with the ethylene receptor in plants. Plant Biol 5:473–480
Sisler EC, Yang SF (1984) Anti-ethylene effects of *cis*-2-butene and cyclic olefins. Phytochemistry 23:2765–2768
Sisler EC, Goren R, Huberman M (1985) Effect of 2,5-norbornadiene on abscission and ethylene production in citrus leaf explants. Physiol Plant 63:114–120
Sisler EC, Blankenship SM, Guest M (1990) Competition of cyclooctenes and cyclooctadienes for ethylene binding and activity in plants. Plant Growth Regul 9:157–164
Sisler EC, Blankenship SM, Fearn JC, Haynes R (1993) Effect of diazocyclopentadiene (DACP) on cut carnations. In: Pech JC, Latché A, Balague C (eds) Cellular and molecular aspects of the plant hormone ethylene. Kluwer, Dordrecht, pp 182–187

Sisler EC, Dupille E, Serek M (1996a) Effect of 1-methylcyclopropene, and methylenecyclopropene on ethylene binding and ethylene action in cut carnations. Plant Growth Regul 18:79–86

Sisler EC, Serek M, Dupille E (1996b) Comparison of cyclopropenes, 1-MCP and 3, 3-dimethylcyclopropene as an ethylene antagonist in plants. Plant Growth Regul 18:169–174

Sisler EC, Serek M, Dupille E, Goren R (1999) Inhibition of ethylene responses by 1-methylcyclopropene and 3-methylcyclopropene. Plant Growth Regul 27:105–111

Sisler EC, Serek M, Roh KA, Goren R (2001) The effect of chemical structure on the antagonism by cyclopropenes of ethylene responses in banana. Plant Growth Regul 33:107–110

Sisler EC, Alwan T, Goren R, Serek M, Apelbaum A (2003) 1-substituted cyclopropenes: effective blocking agents for ethylene action in plants. Plant Growth Regul 40:223–228

Steinite I, Gailite A, Ievinsh G (2004) Reactive oxygen and ethylene are involved in the regulation of regurgitant-induced responses in bean plants. J Plant Physiol 161:191–196

Taiz L, Zeiger E (2002) Plant physiology, 3rd edn. Sinauer Associates, Sunderland, Massachusetts, USA, p 536

Theologis A, Zarembinski TI, Oeller PW, Liang X, Abel S (1992) Modification of fruit ripening by suppressing gene expression. Plant Physiol 100:549–551

Thompson JS, Harlow RL, Whitney F (1983) Copper(I)-olefin complexes: support for the proposed role of copper in the ethylene effect in plants. J Am Chem Soc 105:3522–3527

Thompson JS, Whitney F (1984) Copper(I) complexes with unsaturated small molecules. Preparation and structural characterization of copper(I)-di-2-pyridylamine complexes with olefins, acetylene, and carbon monoxide. Inorg Chem 23:2813–2819

Thompson JS, Swiatek RM (1985) Copper(I) complexes with unsaturated small molecules. Synthesis and properties of monolefin and carbonyl complexes. Inorg Chem 24:110–113

Tian MS, Bowen JH, Bauchot AD, Gong YP, Lallu N (1997) Recovery of ethylene biosynthesis in diazocyclopentadiene (DACP)-treated tomato fruit. Plant Growth Regul 22:73–78

Tian M, Gupta D, Lei XY, Prakash S, Xu C, Fung RWM (2004) Effects of low temperature and ethylene on alternative oxidase in green pepper (*Capsicum annuum* L.). J Hort Sci Biotech 79:493–499

Trinchero GD, Sozzi GO, Covatta F, Fraschina AA (2004) Inhibition of ethylene action by 1-methylcyclopropene extends postharvest life of 'Bartlett' pears. Postharvest Biol Technol 32:193–204

Triola G, Fabrias G, Llebaria A (2001) Synthesis of a cyclopropene analogue of ceramide, a protein inhibitor of dihydroceramide desaturase. Angew Chem Int Edit 40:1960–1962

Van Zhong G, Burns JK (2003) Profiling ethylene-regulated gene expression in *Arabidopsis thaliana* by microarray analysis. Plant Mol Biol 53:117–131

Visser JP, Schipperijn AJ, Lukas J (1973) Platinum(0) complexes of cyclopropenes. J Organomet Chem 47:433–438

Voet-van-Vormizeele J, Groth G (2003) High-level expression of the *Arabidopsis thaliana* ethylene receptor protein ETR1 in *Escherichia coli* and purification of the recombinant protein. Protein Expres Purify 32:89–94

Wang W, Hall AE, O'Malley R, Bleecker AB (2003) Canonical histidine kinase activity of the transmitter domain of the ETR1 ethylene receptor from *Arabidopsis* is not required for signal transduction. Proc Natl Acad Sci USA 100:352–357

Warner HL, Leopold AC (1971) Timing of growth regulator responses in peas. Biochem Biophys Res Commun 44:989–994

Weiler EW (2003) Sensory principles of higher plants. Angew Chem Ger Edit 42:392–411

Woeste KE, Kieber JJ (2000) A strong loss-of-function mutation in RAN1 results in constitutive activation of the ethylene response pathway as well as a rosette-lethal phenotype. Plant Cell 12:443–455

Yamane K, Yamaki Y, Fujishige N (2004) Effects of exogenous ethylene and 1-MCP on ACC oxidase activity, ethylene production and vase life in *Cattleya alliances*. J Jpn Soc Hort Sci 73:128–133

Yokotani N, Tamura S, Nakano R, Inaba A, McGlasson WB, Kubo Y (2004) Comparison of ethylene- and wound-induced responses in fruit of wild-type, *rin* and *nor* tomatoes. Postharvest Biol Technol 32:247–252

2 Ethylene and Plant Growth

Danny Tholen, Hendrik Poorter, Laurentius A.C.J. Voesenek

2.1 Introduction

Plant hormones play an important role in the regulation of growth processes and have profound consequences for plant morphology, development, and carbon gain (Brenner and Cheikh 1995; Mansfield and McAinsh 1995; Reid and Howel 1995). One of these hormones is the gas ethylene. Molecular-genetic analysis has revealed mechanisms responsible for ethylene production, perception, and signal-transduction (Wang et al. 2002; Guo and Ecker 2004). Figure 2.1 summarizes the main regulatory proteins involved in these processes. Cross-talk between components of this pathway and other signal-transduction chains has been identified (Vogel et al. 1998; Bauly et al. 2002; Sharp 2002; León and Sheen 2003; Lorenzo et al. 2003; Pierik et al. 2004), revealing a signal-transduction network that controls numerous aspects of plant growth, development, and responses to external stimuli.

The effect of ethylene on plant growth processes has been investigated by applying both ethylene and ethylene-inhibiting substances. Advances in molecular biology resulted in the availability of several mutant and transgenic plants with impaired ethylene production or perception mechanisms, thus making it possible to study the effect of endogenous ethylene without using ethylene inhibitors, which often have toxic side effects. Most responses to ethylene examined so far are still limited to a particular developmental process (e.g., germination, fruit ripening, pathogen interaction, and senescence), or to elongation growth of individual organs (Abeles et al. 1992). In contrast, this review focuses on the role of endogenous ethylene in the control of whole plant growth.

Plant growth is an important agronomic trait in crops, as well as in wild species growing in their natural habitats. The concept of relative growth rate (RGR) can be used to characterize differences in plant growth. RGR is defined as the increase of mass per unit mass already present per unit time (Blackman 1919; Radford 1967). Fixation of CO_2 by photosynthesis, carbon loss by respiration, as well as uptake of minerals, all contribute to the growth rate of a

---

Utrecht University, Sorbonnelaan 16, Utrecht, 3584 CA The Netherlands

Ethylene Action in Plants
(ed. by N.A. Khan)
© Springer-Verlag Berlin Heidelberg 2006

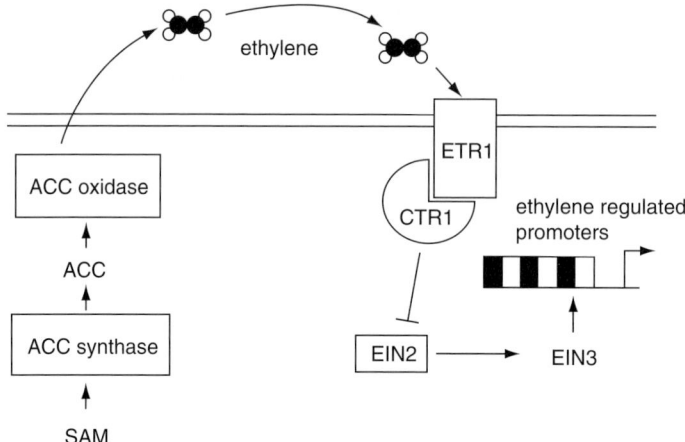

Fig. 2.1. Ethylene production, perception, and signal transduction. Ethylene is formed from the amino acid methionine via S-adenosyl-methionine (SAM), which is converted into 1-aminocyclopropane-1-carboxylic acid (ACC) by ACC synthase. ACC is subsequently hydrolyzed by ACC oxidase (Wang et al. 2002). Ethylene is perceived by a family of membrane bound receptors, including ethylene response 1 (ETR1). When no ethylene is bound the receptor stimulates CTR1 (constitutive triple response 1) which suppresses EIN2 (ethylene insensitive 2) action. EIN2 stimulates the activity of various transcription factors, including EIN3. EIN3 stimulates the transcription of genes involved in ethylene responses (Guo and Ecker 2004)

plant. Therefore, a number of plant traits such as the rate of photosynthesis or leaf morphology, affect the RGR of a plant. The physiological characteristics associated with inherently fast and slow relative growth rates have been extensively investigated (for an overview, see Lambers et al. 1998).

There is little information available about the effects of ethylene on growth in terms of biomass accumulation although some agricultural studies describe the effect of ethylene releasing and inhibiting substances on RGR (Davanso et al. 2003). Even less is known about the role of endogenous ethylene on RGR. In this review we will summarize the known effects of ethylene on leaf expansion and carbon gain, and discuss the different ways by which ethylene can affect whole plant RGR.

2.2 Plant Growth Rate

A frequently used approach to achieve a better understanding of quantitative relationships between the factors underlying whole plant growth rate is a growth analysis such as described by Evans (1972). In such an analysis, RGR is calculated from parameters representing a plant's specific morphological and physiological traits. A mathematical description of this analysis is outlined below.

2.2.1 Plant Growth Analysis

The biomass of a plant at time t (M_t) can be calculated from its biomass at a previous time-point (M_0) as:

$$M_t = M_0 e^{RGR \cdot t}, \quad (2.1)$$

where RGR is the net biomass increase per unit mass already present per day (RGR, mg g^{-1} day^{-1}). RGR can be factorized into three components (Evans 1972). The first component is the leaf area per leaf dry mass or specific leaf area (SLA, m^2 kg^{-1}). The second is the fraction of total biomass allocated to the leaves or leaf mass fraction (LMF, g g^{-1}). The third is the increase of biomass per unit leaf area per day or unit leaf rate (ULR, g m^{-2} day^{-1}). Thus, the formula for RGR can also be written as:

$$RGR = SLA \times LMF \times ULR. \quad (2.2)$$

The SLA is defined as the leaf area (A) divided by the leaf mass (M_L). Leaves that vary in SLA will differ in the amount of photosynthetically active radiation captured per unit of mass. For example, a leaf acclimated to low light conditions generally has a large SLA, and captures more light per unit mass than a leaf with a small SLA. Variation in SLA is of major importance when explaining inherent differences in RGR between species (Poorter and Remkes 1990; Garnier 1992; Atkin et al. 1996). Leaf expansion rate and the net change of mass can affect the SLA of a leaf (Tardieu et al. 1999):

$$SLA_{t + \Delta t} = \frac{A + A/\Delta t}{M_L + M_L/\Delta t}, \quad (2.3)$$

where $\Delta A/\Delta t$ is the expansion rate of the leaves (m^2 day^{-1}), $\Delta M_L/\Delta t$ (in g day^{-1}) is the net outcome of the input of mass (mainly carbohydrates produced by photosynthesis) minus the output of mass from the leaves by respiration and export to the rest of the plant.

LMF is calculated as the leaf mass divided by the plant mass (M):

$$LMF = M_L/M. \quad (2.4)$$

ULR depends on (i) the rate of photosynthesis per unit leaf area (PS_A, mol C fixed m^{-2} leaf area day^{-1}); (ii) the fraction of daily fixed carbon that is not respired but incorporated into the biomass of a plant (FCI, mol C incorporated mol^{-1} C fixed); and (iii) the amount of biomass that can be formed with 1 mol carbon, referred to by the carbon concentration ([C], mol C g^{-1} dry mass). This can be represented as (Poorter 2002):

$$ULR = (PS_A \times FCI)/[C]. \quad (2.5)$$

A change in the SLA of a leaf is a consequence of the expansion rate and the change in leaf mass per unit time (Tardieu et al. 1999, Eq. 2.3).

The net mass change is governed by the input, mainly in the form of photosynthesis, and by the export of carbohydrates and other compounds from the leaves. Plants with a slow rate of leaf expansion are expected to have a smaller SLA compared to plants with a faster leaf expansion. For example, Tardieu et al. (1999) show that a water deficit that decreases expansion rate, but does not affect photosynthesis, results in a smaller SLA in sunflower leaves.

The second factor affecting RGR is the allocation of biomass to leaves (leaf mass fraction, LMF). More allocation of mass to leaves, compared to stems and roots, results in a larger leaf area (provided that SLA remains constant), and thus more light is captured per unit plant mass.

The third factor is the increase in biomass per unit leaf area per day and is called unit leaf rate (ULR). ULR is driven by the carbon fixation in the process of photosynthesis. A part of the fixed carbon is respired by shoots and roots, providing energy for biosynthesis and maintenance; the remaining carbon is combined with nutrients from the soil and incorporated into the structural biomass of the plant. The chemical composition of plants may differ, depending on the amount of minerals combined with carbon from photosynthesis. For example, in a plant that produces a substantial amount of compounds with low carbon content, less carbon is needed to produce one unit of biomass, and this will result in a higher ULR. Accumulation of lipids, lignin, or other carbon-rich components may in turn lead to a low ULR (Poorter and Bergkotte 1992; van Arendonk and Poorter 1994). Environmental factors affecting photosynthesis such as light and CO_2 levels, or genetic factors such as a decreased level of Rubisco in transgenic plants, can have dramatic effects on the ULR. In such cases, growth rate is often maintained to a certain extent by compensatory changes in SLA (Quick et al. 1991; Poorter and Nagel 2000).

2.3 Ethylene and Leaf Expansion

Leaf expansion is the result of both cell production and cell expansion (Beemster et al. 2003). There are few reports of ethylene stimulating cell division (Zobel and Roberts 1978; Raz and Koornneef 2001). However, a large amount of data is available describing the role of ethylene in cell expansion (Abeles et al. 1992). Large concentrations of externally applied ethylene generally inhibit elongation (Abeles et al. 1992). One of the most well-known elongation inhibiting effects of ethylene is the 'triple response' of seedlings grown in darkness, and was first discovered in pea seedlings exposed to ethylene (Neljubow 1901). This effect includes inhibition of stem elongation, radial swelling of the stem, and agravitropic growth (Knight and Crocker 1913). In *Arabidopsis* seedlings, the triple response is characterized by a shortened and thickened hypocotyl, an inhibition of root elongation, and the formation of an exaggerated apical hook (Guzmán and Ecker 1990). In addition to these findings, constitutively increased ethylene levels also reduce elongation in adult plants. For example, stems of transgenic tobacco with an elevated ethylene production level have shorter internodes, resulting in reduced plant height (Romano et al. 1993; Knoester et al. 1997). The *Arabidopsis* ethylene-transduction mutant *ctr1* exhibits a phenotype similar to a plant treated with high ethylene concentrations, and has a dramatically reduced stature and unexpanded leaves (Kieber et al. 1993; Ecker 1995). The smaller leaf area is correlated with a smaller cell size in the *ctr1* plants, similar to the reduction in cell size caused by treatment with ethylene (Kieber et al. 1993).

The reduction of cell expansion by ethylene is consistent with the finding that leaf area is increased in mutant or transgenic plants impaired in ethylene perception. Larger rosette diameters have been observed in ethylene-insensitive *Arabidopsis* mutants (Guzmán and Ecker 1990; Ecker 1995). In more detail, the total leaf area of rosette leaves of ethylene-insensitive *Arabidopsis* mutants (*etr1–1* and *ers1*) was reported to be 25–50% larger than those of wild-type plants (Bleecker et al. 1988; Hua et al. 1995). The increase in total leaf area was attributed to an increased cell expansion in ethylene-insensitive plants (Hua et al. 1995). A greater leaf area in ethylene-insensitive plants may indicate that even the low endogenous ethylene concentration in the wild type has an inhibiting effect on leaf expansion. However, it is also possible that the larger total leaf area in ethylene-insensitive plants is the result of an increased leaf longevity. Leaves of ethylene-insensitive plants continue to expand over a long time period, whereas the leaves of wild-type plants of the same age have a much shorter expansion period (Grbić and Bleecker 1995). Alternatively, the relatively larger leaf area of ethylene-insensitive *Arabidopsis* mutants grown in sterile tissue-culture containers may be explained by a reduction of the leaf area of the wild type as a result of ethylene accumulation (Tholen et al. 2004). We showed that under well-ventilated conditions ethylene-insensitive genotypes of *Arabidopsis*, tobacco and petunia had no larger total leaf area compared to normally ethylene-sensitive control plants (Fig. 2.2, Tholen et al. 2004).

In contrast with the finding that ethylene treatment inhibits elongation growth are reports that low concentrations (i.e., below 0.1 μL L^{-1}) can stimulate leaf expansion (Lee and Reid 1997; Fiorani et al. 2002), stem elongation (Emery et al. 1994; Suge et al. 1997; Pierik et al. 2003), hypocotyl elongation (Smalle et al. 1997), and root elongation (Konings and Jackson 1979). To explain these differential responses to ethylene, a biphasic model was suggested (Konings and Jackson 1979; Lee and Reid 1997), with low levels promoting and high levels inhibiting cell expansion. In aquatic and semi-aquatic species, the concentration range promoting growth may be wider in order to allow the observed rapid elongation growth under submerged conditions that is initiated by ethylene levels of up to 10 μL L^{-1} (Voesenek and Blom 1999). For example, in *Oryza sativa* and *Rumex*

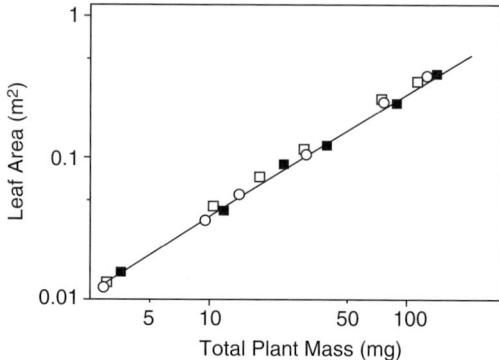

Fig. 2.2. Total leaf area of *Arabidopsis* thaliana wild-type (*closed squares*), ethylene-insensitive *etr1-1* (*open squares*), and ethylene-insensitive *ein2-1* (*open circles*) mutants plotted against total plant mass. Plants were grown in a climate room (T=20°C, RH=65%, 200 µmol m^{-2} s^{-1} PAR for 16 h) on hydroponics and harvested six times during a 2-week growth period. Each data point is the average of six plants. A trend line is shown for the wild-type plants. The leaf area of the ethylene-insensitive mutants was not significantly different from that of the wild type ($P<0.05$)

palustris, ethylene induces a rapid elongation of internodes or petioles, which enables the leaf blade to escape from the low oxygen environment under water (Kende et al. 1998; Voesenek and Blom 1999). Ethylene accumulates in plants that are completely submerged as a result of the slower diffusion rate of gases in water compared to air (Jackson 1985). The high ethylene concentration promotes organ elongation in these systems through direct action associated with changes in the balance of gibberellic acid and abscisic acid (Hoffmann-Benning and Kende 1992; Rijnders et al. 1997; Benschop 2004).

The network of signal-transduction components that affect organ expansion includes gibberellic acid, auxin, and ethylene. A key component that integrates signals from all these hormones are a group of proteins now called DELLA proteins (Achard et al. 2003; Fu and Harberd 2003; Harberd 2003). It has been shown that at least a part of the regulatory effects of ethylene on expansion is mediated by DELLA proteins, which act as repressors of elongation (Achard et al. 2003). The reduction in elongation by ethylene is thought to be the result of reduced cell-elongation induced by a reorientation of cortical microtubuli (Roberts et al. 1985; Shibaoka 1994). However, a more recent work by Le et al. (2004) could not confirm this reorientation of microtubuli in *Arabidopsis* roots. Another mechanism by which ethylene, possibly via DELLA protein action, can affect cell expansion is by regulation of the expression or activity of cell-wall loosening proteins such as expansins (Cosgrove et al. 2002; Vreeburg et al. 2005).

2.4 Ethylene and Photosynthesis

A mature leaf will undergo senescence accompanied by a decline in photosynthesis, followed by reallocation of nitrogen to other parts of the plant (Gan and Amasino 1997). Temporary treatment of leaves of adult plants with high ethylene concentrations generally causes chlorophyll loss, senescence, and leaf abscission (Kays and Pallas 1980; Bleecker et al. 1988; Abeles et al. 1992). Interestingly, it has been found that ethylene-mediated pathways leading to leaf senescence in *Arabidopsis* depend on age-related factors, and ethylene can only induce senescence after the leaves reach a certain developmental stage (Grbić and Bleecker 1995; Weaver et al. 1998). This last observation may explain why constitutive ethylene responding mutants (*ctr1*) and wild-type plants grown continuously in ethylene do not show increased senescence during the vegetative growth phase (Kieber et al. 1993).

Ethylene may also affect the process of photosynthesis via senescence-independent pathways. For example, there have been a number of studies of *Brassica juncea* showing that a range of ethylene concentrations can promote the rate of photosynthesis (Subrahmanyam and Rathore 1992; Khan 2004), probably as a result of an increased stomatal conductance (Khan 2004). Conflicting results have been reported in the literature concerning the effect of ethylene on stomatal conductance (Gunderson and Taylor 1991; Kamaluddin and Zwiazek 2002). The effect of ethylene on photosynthesis seems to be concentration dependent (Khan 2004). In addition, we recently found that stomatal conductance was lower in ethylene-insensitive *Arabidopsis* mutants, but higher in ethylene-insensitive tobacco when compared to ethylene-sensitive controls (D. Tholen, unpublished research). This indicates that the effect of ethylene on stomatal regulation, and thus photosynthesis, may differ between species.

In addition to effects on stomatal conductance, ethylene may play a role in the regulation of photosynthesis via its role in sugar sensing. One of the mechanisms that control photosynthetic activity and gene expression is a down-regulation of Calvin-cycle enzymes by carbohydrate end-products (Paul and Pellny 2003). Recent studies on the interaction between ethylene, abscisic acid (ABA), and sugar sensing suggest that ethylene may play a role in the regulation of photosynthetic gene expression (León and Sheen 2003). Ethylene-insensitive plants appear to be more sensitive to endogenous glucose levels, while application of an ethylene precursor decreases a plant's sensitivity to glucose (Zhou et al. 1998). A greater sensitivity of ethylene- insensitive plants to endogenous glucose may result in a slower rate of photosynthesis in these plants. In support of this view is the observation by Grbić and Bleecker (1995) that young leaves of ethylene-insensitive *etr1–1* mutants have less chlorophyll and a lower rate of carboxylation per unit area than wild-type *Arabidopsis* leaves. Recently, we found that ethylene-insensitive tobacco plants have a lower photosynthetic capacity as a result of a lower Rubisco expression. The effect was linked to an

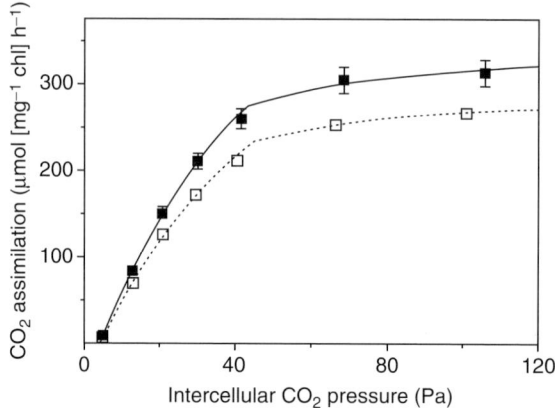

Fig. 2.3. CO_2-response curves of the rate of photosynthesis in ethylene-sensitive wild-type *Arabidopsis* thaliana (*solid squares*) and ethylene-insensitive (*open squares*) etr1-1 mutants. Lines were calculated using the biochemical model of photosynthesis as described by Farquhar and von Caemmerer (1982). Error bars larger than the plot symbol are shown and indicate standard error (n=8)

increased sensitivity to endogenous glucose levels (D. Tholen, T.L. Pons, L.A.C.J. Voesenek, H. Poorter, unpublished results). In Fig. 2.3 we have shown the response of photosynthesis to different CO2 partial pressures at saturating light levels. From these data it can be calculated that the *Arabidopsis etr1-1* mutant has a significantly lower carboxylation capacity (17%, P<0.05), confirming the results found in the ethylene-insensitive tobacco plants.

Increased ethylene production may be central in promoting growth under elevated ambient carbon dioxide levels (Seneweera et al. 2003) to overcome the negative effect of high endogenous glucose concentrations on photosynthesis. This may explain why ethylene production in the light is increased in response to high CO_2 concentrations (Bassi and Spencer 1982). We suggest that low ethylene concentrations can have a promoting effect on photosynthesis under circumstances where carbohydrate-repression plays a limiting role, such as under high light levels or increased atmospheric CO_2 concentrations.

In summary, it is unlikely that the low endogenous ethylene concentration has a negative impact on photosynthetic performance. The suppressing effect of ethylene on glucose sensitivity could even be a mechanism by which growth is maintained under conditions where sugars accumulate such as in an active photosynthesizing leaf.

2.5 Ethylene and Growth under Optimal Conditions

Ethylene plays an important role in the regulation of leaf expansion; high ethylene concentrations generally inhibit expansion, whereas in some cases, low concentrations can stimulate leaf expansion. As was previously described (Eq. 2.3), a change in leaf expansion by ethylene may result in a difference in SLA. Since ethylene treatment normally inhibits expansion, the absence of ethylene or an impaired ability to perceive ethylene may increase leaf expansion. This does not necessarily result in a larger total leaf area per plant because of the lower total biomass of ethylene-insensitive plants (Tholen et al. 2004). However, we did find an increase in the total leaf area per unit biomass in ethylene-insensitive genotypes when compared to their ethylene-sensitive counterparts (Tholen et al. 2004).

Large concentrations of externally applied ethylene induce senescence, and this can lead to a slower ULR as a result of the breakdown of the photosynthetic machinery. A low concentration of ethylene might also stimulate photosynthesis, and thus ULR, through an increase of stomatal conductance. In ethylene-insensitive plants photosynthesis (and ULR) was generally lower than in the corresponding ethylene-sensitive genotypes (Tholen et al. 2004).

Ethylene can be involved in the partitioning of biomass over different plant organs, but to date, no studies have addressed this in depth. Clark et al. (1999) found that ethylene insensitive tomato plants had increased below ground biomass, but fewer above ground adventitious roots. However, treatment of *Phaseolus vulgaris* with the ethylene inhibitor AVG had no significant effect on LMF (Borch et al. 1999).

To our knowledge, no reports exist describing the response of RGR to a controlled application of ethylene. Davanso et al. (2003) showed that application of the ethylene-releasing substance ethephon to *Tabebuia avellanedae* plants resulted in a reduced ULR and RGR. RGR may have been lowered as a result of a high leaf abscission induced by the relatively large amount of ethylene released by ethephon. In addition, these authors also showed that ULR and RGR were reduced as a result of the application of silver nitrate, an ethylene perception inhibitor. Recently, we showed that ethylene-insensitive plants have a higher SLA but a slower ULR, resulting in a RGR that was indistinguishable from ethylene-sensitive plants (Tholen et al. 2004). Therefore we suggest that although endogenous ethylene is able to affect the underlying mechanisms that determine RGR, it is not required for controlling growth under optimal conditions.

2.6 Ethylene and Growth under Limiting Environmental Conditions

Ethylene has long been recognized as a hormone that controls plant responses under growth-limiting conditions or stress (for reviews see Abeles et al. 1992; Morgan and Drew 1997). An important environmental factor with an effect on growth is light intensity. Plants have the ability to acclimate to a sub-optimal light environment by increasing SLA (Poorter and Nagel 2000). When growing at high densities with competition for light, plants show several shade-avoidance responses, including enhanced elongation, elevated leaf angles, and early flowering (Smith 2000). Interestingly, it was found that ethylene controls the timing of these shade-avoidance responses (Pierik et al. 2003). Moreover, recent evidence shows that ethylene-sensitive tobacco plants exhibit shade-avoidance responses when blue light is omitted from the light spectrum, whereas the ethylene-insensitive plants are unresponsive to the absence of blue light (Pierik et al. 2004). These results suggest that the ethylene-transduction pathway interacts with light perception and that ethylene is required to maintain growth during competition for light.

A limited nutrient availability can severely constrain plant growth in natural environments (Berendse and Elberse 1990). It has been shown that ethylene synthesis and responsiveness is affected in response to nutrient deficiency (Lynch and Brown 1997). For instance, nitrogen deficiency and low pH increased the ethylene production in wheat seedlings (Tari and Szén 1995). Borch et al. (1999) showed that phosphorus-deficient roots produce twice as much ethylene as phosphorus-sufficient roots. Phosphorus or nitrogen starvation reduced ethylene production, but increased ethylene responsiveness in maize roots (Drew et al. 1989; He et al. 1992). Borch et al. (1999) suggested that root responses to phosphorus deficiency are controlled by ethylene sensitivity and production. Along similar lines, Schmidt et al. (2000) showed evidence for the hypothesis that ethylene (and possibly auxin) mediates iron-deficiency-induced root-hair formation. In addition to the effects on ethylene production or sensitivity, nutrients may also affect the way a plant responds to ethylene. For example, in *Arabidopsis* hypocotyls, ethylene stimulated elongation only when the plants were grown on a medium with low nitrogen and phosphate concentrations (Smalle et al. 1997; D. Tholen, unpublished results).

Drought stress is a major factor hampering crop yield. Under drought stress shoot elongation can be inhibited, even before water potential is decreased in aerial plant parts (Saab and Sharp 1989), suggesting that signals emanating from the root may affect shoot growth. Increased root production of abscisic acid (ABA) and the ethylene precursor ACC (Gómez-Cadenas et al. 1996) might be the responsible signals. Although these authors concluded that ABA stimulates ACC synthesis, later work by Spollen et al. (2000) suggested that increased levels of endogenous ABA are required to prevent excess ethylene production, which has a negative effect on root elongation. Transgenic maize with a lower level of ACC-synthase has a reduced rate of ethylene production, and a faster rate of photosynthesis under drought stress (Young et al. 2004).

Similar results were found for ethylene production and salt stress. The rate of ethylene production was increased in salt-stressed tomato and cucumber, promoting senescence in these plants (Feng and Barker 1992; Helmy et al.

1994). Loss of chlorophyll under salt stress has been reported due to ethylene-induced reduction in intermediates of chlorophyll biosynthesis (Khan 2003). Salt-tolerant rice species synthesize more ethylene compared to less tolerant varieties (Lutts et al. 1996). In tobacco, transcript levels of a putative ethylene-receptor homologue are increased under drought and salt stress (Zhang et al. 2001), suggesting that not only production, but also sensitivity to ethylene can be affected by these environmental conditions. Ethylene-insensitive tobacco plants, grown for 2 weeks under severe drought stress showed no differences in RGR compared to ethylene-sensitive plants. However, salt-stressed ethylene-insensitive tobacco plants showed a reduction in RGR that was not apparent in normally ethylene-sensitive tobacco (D. Tholen, H. Poorter, L.A.C.J. Voesenek, unpublished results). This suggests that a functional ethylene-response pathway is needed to maintain growth under salt-stressed conditions.

Together, these data show that ethylene is involved in a number of stress responses that occur in natural environments and can significantly affect whole plant growth under adverse circumstances.

2.7 Ethylene: Effects on Plant Growth?

The role of normally occurring levels of endogenous ethylene in determining the rate of plant growth is a long standing question. Most of the work has focused on tissue- or cell-specific responses, and little is known about the effects of ethylene at the whole plant level. In this review, we examined the effect of ethylene on growth in terms of biomass accumulation. As explained above, ethylene treatment will generally lower the RGR of a plant, especially if high concentrations are used. No increase in RGR was reported for several ethylene-insensitive plants, suggesting that the normally present, endogenous ethylene concentrations are not inhibiting RGR. Furthermore, an increasing body of literature reveals a stimulating effect of low ethylene concentrations on both leaf expansion and photosynthesis.

Under stress conditions, ethylene production rates are generally strongly increased. In these circumstances, ethylene often acts as a signal that initiates a response to the stress. It can be expected that the high ethylene levels initially produced as a consequence of the stress inhibit growth and will reduce RGR. However, an adequate response to the stress may be beneficial to growth and survival in the long run.

Acknowledgements: We thank Dr. Fabio Fiorani and Prof. Mike Jackson for their helpful comments on an earlier version of this manuscript. This work was supported by the Earth and Life Sciences Foundation, which is subsidized by the Netherlands Organization for Scientific Research (NWO; grant no. 805.33.463) and by the NWO PIONIER grant no. 800.84.470 to L.A.C.J.V.

References

Abeles FB, Morgan PW, Saltveit ME (1992) Ethylene in plant biology. Academic, New York

Achard P, Vriezen WH, Van Der Straeten D, Harberd NP (2003) Ethylene regulates *Arabidopsis* development via the modulation of DELLA protein growth repressor function. Plant Cell 15:2816–2825

Atkin OK, Botman B, Lambers H (1996) The causes of inherently slow growth in alpine plants: an analysis based on the underlying carbon economies of alpine and lowland *Poa* species. Funct Ecol 10:698–707

Bassi PK, Spencer MS (1982) Effect of carbon dioxide and light on ethylene production in intact sunflower plants. Plant Physiol 69:1222–1225

Bauly J, Roux C, Dargeviciute A, Perrot-Rechenmann C (2002) Identification of a novel molecular marker for auxin and ethylene cross-talk from tobacco seedlings. Plant Physiol Biochem 40:803–811

Beemster GTS, Fiorani F, Inze D (2003) Cell cycle: the key to plant growth control. Trends Plant Sci 8:154–158

Benschop JJ (2004) The role of abscisic acid in ethylene-induced elongation. Thesis, Utrecht University, Utrecht

Berendse F, Elberse WT (1990) Competition and nutrient availability in heathlands and grass ecosystems. In: Grace JB, Tilman D (eds) Perspectives on plant competition. Academic, San Diego, pp 93–116

Blackman VH (1919) The compound interest law and plant growth. Ann Bot 23:353–360

Bleecker AB, Estelle MA, Somerville C, Kende H (1988) Insensitivity to ethylene conferred by a dominant mutation in *Arabidopsis thaliana*. Science 141:1086–1087

Borch K, Bouma TJ, Lynch JP, Brown KM (1999) Ethylene: a regulator of root architectural responses to soil phosphorus availability. Plant Cell Environ 22:425–431

Brenner ML, Cheikh N (1995) Hormones in photosynthate partitioning and seed filling. In: Davies PJ (ed) Plant hormones. Kluwer, Dordrecht, pp 649–670

Clark DG, Gubrium EK, Barrett JE, Nell TA, Klee HJ (1999) Root formation in ethylene-insensitive plants. Plant Physiol 121:53–59

Cosgrove DJ, Li LC, Cho HT, Hoffmann-Benning S, Moore RC, Blecker D (2002) The growing world of expansins. Plant Cell Physiol 43:1436–1444

Davanso VM, Medri ME, de Souza LA, Colli S (2003) *Tabebuia avellanedae* Lor. ex Griseb. (Bignoniaceae) submitted at the flooding and the "Ethrel" and silver nitrate application. Braz Arch Biol Technol 46:57–64

Drew MC, He CJ, Morgan PW (1989) Decreased ethylene biosynthesis, and induction of aerenchyma, by nitrogen- or phosphate-starvation in adventitious roots of *Zea mays* L. Plant Physiol 91:266–271

Ecker JR (1995) The ethylene signal transduction pathway in plants. Science 268:667–675

Evans GC (1972) The quantitative analysis of plant growth. Blackwell Scientific Publ, Oxford

Emery RJN, Reid DM, Chinnappa CC (1994) Phenotypic plasticity of stem elongation in two ecotypes of *Stellaria longipes*: the role of ethylene response to wind. Plant Cell Environ 17:691–700

Farquhar GD, von Caemmerer S (1982) Modelling of photosynthetic response to environmental conditions. In: Lange OL, Nobel PS, Osmond CB, Ziegler H (eds) Encyclopedia of plant physiology, new series, vol 12B. Physiological plant ecology II. Springer, Berlin Heidelberg New York, pp 549–587

Feng J, Barker AV (1992) Ethylene evolution and ammonium accumulation by tomato plants under water and salinity stress. J Plant Nutr 15:2471–2490

Fiorani F, Bögemann GM, Visser EJW, Lambers H, Voesenek LACJ (2002) Ethylene emission and responsiveness to applied ethylene vary among *Poa* species that inherently differ in leaf elongation rates. Plant Physiol 129:1382–1390

Fu X, Harberd NP (2003) Auxin promotes *Arabidopsis* root growth by modulating gibberellin response. Nature 421:740–743

Gan S, Amasino RM (1997) Making sense of senescence: molecular genetic regulation and manipulation of leaf senescence. Plant Physiol 113:313–319

Garnier E (1992) Growth analysis of congeneric annual and perennial grass species. J Ecol 80:665–675

Gómez-Cadenas A, Tadeo FR, Talon M, Primo-Millo E (1996) Leaf abscission induced by ethylene in water stressed intact seedlings of (*Citrus reshni* Hort. ex Tan.) requires previous abscisic acid accumulation in roots. Plant Physiol 112:401–408

Grbić V, Bleecker AB (1995) Ethylene regulates the timing of leaf senescence in *Arabidopsis*. Plant J 8:595–602

Gunderson CA, Taylor GE Jr (1991) Ethylene directly inhibits foliar gas exchange in *Glycine max*. Plant Physiol 95:337–339

Guo H, Ecker JR (2004) The ethylene signaling pathway: new insights. Curr Opin Plant Biol 7:40–49

Guzmán P, Ecker JR (1990) Exploiting the triple response of *Arabidopsis* to identify ethylene-related mutants. Plant Cell 2:513–523

Harberd NP (2003) Relieving DELLA restraint. Science 299:1853–1854

He CJ, Morgan PW, Drew MC (1992) Enhanced sensitivity to ethylene in nitrogen-starved or phosphate-starved roots of *Zea mays* L. during aerenchyma formation. Plant Physiol 98:137–142

Helmy YH, El Abd SO, Abou Hadid AF, El Beltag U, El Betagy AS (1994) Ethylene production from tomato and cucumber plants under saline conditions. Egypt J Hort 21:153–160

Hoffmann-Benning S, Kende H (1992) On the role of abscisic acid and gibberellin in the regulation of growth in rice. Plant Physiol 99:1156–1161

Hua J, Chang C, Sun Q, Meyerowitz EM (1995) Ethylene insensitivity conferred by *Arabidopsis* ERS gene. Science 269:1712–1714

Jackson MB (1985) Ethylene and responses of plants to soil waterlogging and submergence. Annu Rev Plant Physiol 36:145–174

Kamaluddin M, Zwiazek JJ (2002) Ethylene enhances water transport in hypoxic aspen. Plant Physiol 128:962–969

Kays SJ, Pallas JE Jr (1980) Inhibition of photosynthesis by ethylene. Nature 285:51–52

Kende H, van der Knaap E, Cho HT (1998) Deepwater rice: a model plant to study stem elongation. Plant Physiol 118:1105–1110

Khan NA (2003) NaCl-inhibited chlorophyll synthesis and associated changes in ethylene evolution and antioxidative enzyme activities in wheat. Biol Plant 47:437–440

Khan NA (2004) An evaluation of the effects of exogenous ethephon: an ethylene releasing compound, on photosynthesis of mustard (*Brassica juncea*) cultivars that differ in photosynthetic capacity. BMC Plant Biol 4:21

Kieber JJ, Rothenberg M, Roman G, Feldmann KA, Ecker JR (1993) CTR1, a negative regulator of the ethylene response pathway in *Arabidopsis*, encodes a member of the Raf family of protein kinases. Cell 72:427–441

Knight LI, Crocker W (1913) Toxicity of smoke. Bot Gaz 55:337–371

Knoester M, Linthorst HJM, Bol JF, van Loon LC (1997) Modulation of stress-inducible ethylene biosynthesis by sense and antisense gene expression in tobacco. Plant Sci 126:173–183

Konings H, Jackson MB (1979) A relationship between rates of ethylene production by roots and the promoting or inhibiting effects of exogenous ethylene and water on root elongation. Z Pflanzenphysiol 92:385–397

Lambers H, Poorter H, van Vuuren MMI (1998) Inherent variation in plant growth: physiological mechanisms and ecological consequences. Backhuys Publ, Leiden

Le J, Vandenbussche F, van der Straeten D, Verbelen JP (2004) Position and cell type-dependent microtubule reorientation characterizes the early response of the *Arabidopsis* root epidermis to ethylene. Physiol Plant 121:513–519

Lee S, Reid D (1997) The role of endogenous ethylene in the expansion of *Helianthus annuus* leaves. Can J Bot 75:501–509

León P, Sheen J (2003) Sugar and hormone connections. Trends Plant Sci 8:110–116

Lorenzo O, Piqueras R, Sánchez-Serrano JJ, Solano R (2003) Ethylene Response Factor1 integrates signals from ethylene and jasmonate pathways in plant defense. Plant Cell 15:165–178

Lutts S, Kinet JM, Bouharmont J (1996) NaCl-induced senescence in leaves of rice (*Oryza sativa* L.) cultivars differing in salinity resistance. Ann Bot 78:389–398

Lynch JP, Brown KM (1997) Ethylene and plant responses to nutritional stress. Physiol Plant 100:613–619

Mansfield TA, McAinsh MR (1995) Hormones as regulators of water balance. In: Davies PJ (ed) Plant hormones. Kluwer, Dordrecht, pp 598–616

Morgan PW, Drew MC (1997) Ethylene and plant response to stress. Physiol Plant 100:620–630

Neljubow DN (1901) Über die horizontale Nutation der Stengel von *Pisum sativum* und einiger anderen Pflanzen. Pflanzen Beitr Bot Zentralbl 10:128–139

Paul MJ, Pellny TK (2003) Carbon metabolite feedback regulation of leaf photosynthesis and development. J Exp Bot 54:539–547

Pierik R, Visser EJW, de Kroon H, Voesenek LACJ (2003) Ethylene is required in tobacco to successfully compete with proximate neighbours. Plant Cell Environ 26:1229–1234

Pierik R, Whitelam GC, Voesenek LACJ, de Kroon H, Visser EJW (2004) Canopy studies on ethylene-insensitive tobacco identify ethylene as a novel element in blue light and plant-plant signaling. Plant J 38:310–319

Poorter H (2002) Plant growth and carbon economy. Encyclopedia of Life Sciences, www.els.net. Macmillan Publ, Nature Publ Group, London

Poorter H, Bergkotte M (1992) Chemical composition of 24 wild species differing in relative growth rate. Plant Cell Environ 15:221–229

Poorter H, Nagel O (2000) The role of biomass allocation in the growth response of plants to different levels of light, CO_2, nutrients and water: a quantitative review. Aust J Plant Physiol 27:595–607

Poorter H, Remkes C (1990) Leaf area ratio and net assimilation rate of 24 wild species differing in relative growth rate. Oecologia 83:553–559

Quick WP, Schurr U, Fichtner K, Schulze ED, Rodermel SR, Bogorad L, Stitt M (1991) The impact of decreased Rubisco on photosynthesis, growth, allocation and storage in tobacco plants which have been transformed with antisense *rbcS*. Plant J 1:51–58

Radford PJ (1967) Growth analysis formulae: their use and abuse. Crop Sci 7:171–175

Raz V, Koornneef M (2001) Cell division activity during apical hook development. Plant Physiol 125:219–226

Reid JB, Howel H (1995) Hormone mutants and plant development. In: Davies PJ (ed) Plant hormones. Kluwer, Dordrecht, pp 598–616

Rijnders JGHM, Yang YY, Kamiya Y, Takahashi N, Barendse GWM, Blom CWPM, Voesenek LACJ (1997) Ethylene enhances gibberellin levels and petiole sensitivity in flooding-tolerant *Rumex palustris* but not in flooding-intolerant *R. acetosa*. Planta 203:20–25

Roberts IN, Lloyd CW, Roberts K (1985) Ethylene-induced microtubule reorientation: mediation by helical arrays. Planta 164:439–447

Romano C, Cooper M, Klee HJ (1993) Uncoupling auxin and ethylene effects in transgenic tobacco and *Arabidopsis* plants. Plant Cell 5:181–189

Saab IN, Sharp RE (1989) Non-hydraulic signals from maize roots in drying soil: inhibition of leaf elongation but not stomatal conductance. Planta 179:466–474

Schmidt W, Tittel J, Schikora A (2000) Role of hormones in the induction of Fe deficiency responses in *Arabidopsis* roots. Plant Physiol 122:1109–1118

Seneweera S, Aben SK, Basra AS, Jones B, Conroy JP (2003) Involvement of ethylene in the morphological and developmental response of rice to elevated atmospheric CO_2 concentrations. Plant Growth Regul 39:143–153

Sharp RE (2002) Interaction with ethylene: changing views on the role of abscisic acid in root and shoot growth responses to water stress. Plant Cell Environ 25:211–222

Shibaoka H (1994) Plant hormone-induced changes in the orientation of cortical microtubules: alterations in the cross-linking between microtubules and the plasma membrane. Annu Rev Plant Physiol Plant Mol Biol 45:527–544

Smalle J, Haegman M, Kurepa J, van Montagu M, van der Straeten D (1997) Ethylene can stimulate *Arabidopsis* hypocotyl elongation in the light. Proc Natl Acad Sci USA 94:2756-2761

Smith H (2000) Phytochromes and light signal perception by plants: an emerging synthesis. Nature 407:585-591

Spollen WG, LeNoble ME, Samuels TD, Bernstein N, Sharp RE (2000) Abscisic Acid accumulation maintains maize primary root elongation at low water potentials by restricting ethylene production. Plant Physiol 122:967-976

Suge H, Nishizawa T, Takahashi H, Takeda K (1997) Phenotypic plasticity of internode elongation stimulated by deep-seedling and ethylene in wheat seedlings. Plant Cell Environ 20:961-964

Subrahmanyam D, Rathore VS (1992) Influence of ethylene on carbon-14 labelled carbon dioxide assimilation and partitioning in mustard. Plant Physiol Biochem 30:81-86

Tardieu F, Granier C, Muller B (1999) Modelling leaf expansion in a fluctuating environment: are changes in specific leaf are a consequence of changes in expansion rate? New Phytol 143:33-43

Tari I, Szén L (1995) Effect of nitrite and nitrate nutrition on ethylene production by wheat seedlings. Acta Phytopathol Entomol Hung 30:99-104

Tholen D, Voesenek LACJ, Poorter H (2004) Ethylene insensitivity does not increase leaf area or relative growth rate in *Arabidopsis thaliana, Nicotiana tabacum* and *Petunia x hybrida*. Plant Physiol 134:1803-1812

Van Arendonk JJCM, Poorter H (1994) The chemical composition and anatomical structure of leaves of grass species differing in relative growth rate. Plant Cell Environ 17:963-970

Voesenek LACJ, Blom CWPM (1999) Stimulated shoot elongation: a mechanism of semiaquatic plants to avoid submergence stress. In: Lerner HR (ed) Plant responses to environmental stresses: from phytohormones to genome reorganization. Dekker, New York, pp 431-448

Vogel JP, Schuerman P, Woeste K, Brandstatter I, Kieber JJ (1998) Isolation and characterization of *Arabidopsis* mutants defective in the induction of ethylene biosynthesis by cytokinin. Genetics 149:417-427

Vreeburg RAM, Benschop JJ, Peeters AJM, Colmer TD, Ammerlaan AHM, Staal M, Elzenga TM, Staals RHJ, Darley CP, McQueen-Mason SJ, Voesenek LACJ (2005) Ethylene regulates fast apoplastic acidification and expansin A transcription during submergence-induced petiole elongation in *Rumex palustris*. Plant J 43:597-610

Wang KL, Li H, Ecker JR (2002) Ethylene biosynthesis and signaling networks. Plant Cell 14:131-151

Weaver LM, Gan S, Quirino B, Amasino RM (1998) A comparison of the expression patterns of several senescence-associated genes in response to stress and hormone treatment. Plant Mol Biol 37:455-469

Young TE, Meeley RB, Gallie DR (2004) ACC synthase expression regulates leaf performance and drought tolerance in maize. Plant J 40:813-825

Zhang JS, Xie C, Shen YG, Chen SY (2001) A two-component gene (*NTHK1*) encoding a putative ethylene-receptor homolog is both developmentally and stress regulated in tobacco. Theo Appl Genet 102:815-824

Zhou L, Jang J, Jones T, Sheen J (1998) Glucose and ethylene signal transduction crosstalk revealed by an *Arabidopsis* glucose-insensitive mutant. Proc Natl Acad Sci USA 95:10294-10299

Zobel RW, Roberts LW (1978) Effects of low concentration of ethylene on cell division and cytodifferentiation in lettuce pit explants. Can J Bot 56:987-990

3 Ethylene and Leaf Senescence

ANTONIO FERRANTE[1], ALESSANDRA FRANCINI[2]

3.1 Introduction

Leaf senescence is currently considered the last stage of leaf development and is a genetically programmed process, highly regulated with recycling of reserves from the leaves to other storage organs (seeds, trunk, branches). Leaves of deciduous plants perceive the changing of the seasons in autumn and activate all the processes for the preparation for winter. The signs that initiate natural leaf senescence are the photosynthesis reduction and shortening of photoperiod (Smart 1994; Yoshida 2003). The leaves of evergreens have a turnover regulated by an internal clock that is correlated with the photosynthetic machinery health status. Leaf cells undergo an organized self-destruction process that involves protein degradation (Lutts et al. 1996), and altered turnover, nucleic acid degradation, lipid degradation (Buchanan-Wollaston 1997; Thompson et al. 1998; Buchanan-Wollaston et al. 2003), membranes disruption (Trippi and Thiamann 1983), and leaf pigments breakdown (Matile et al. 1997; Fang et al. 1998). The metabolites and nutrients are reallocated in storage organs such as branches and trunk in deciduous trees, which will use the reserve for making a new leaf area in spring, or in seeds in annual species that will use the reserve for growing a new plant the next year (Noodén and Guiamét 1989; Gan and Amasino 1995). The leaf senescence is activated at the molecular level when leaves are fully expanded (Fig. 3.1). At this stage, many processes are induced and many others are turned off.

The leaf senescence can be subdivided into three phases. The first is named the initiation phase, which may be induced by hormones, environment, age or pathogens. At the physiological level, the metabolic thresholds are crossed, the redox state is altered, and the signalling cascades are activated. The second phase can be influenced by hormones and environment. During this stage, the leaf cells are reorganizing for responding to the degeneration by activation of savage pathways, shifting the metabolism from autotrophic to heterotrophic, detoxification and reversible organelle

---

[1] Dep. Produzione Vegetale, University of Milan, via Celoria 2, I-20133 Italy

[2] Dep. Coltivazione e Difesa delle Specie Legnose "G. Scaramuzzi", University of Pisa, via del Borghetto 80, I-56100, Italy

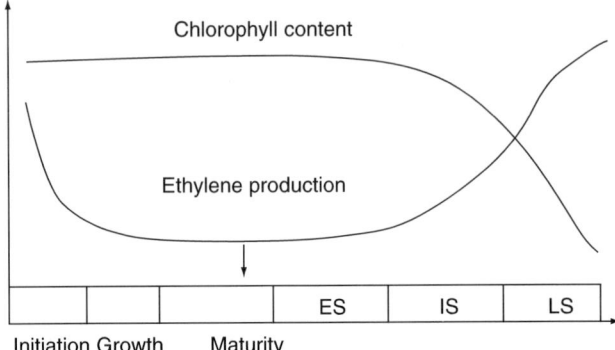

Fig. 3.1. Schematic graph of the leaf development, ethylene production, and chlorophyll content trends. The *arrow* represents the induction of the earliest senescence associate genes that initiate the senescence cascade. *ES* early senescence, *IS* intermediate senescence, *LS* leaf senescence

redifferentiation. The last phase of leaf senescence is characterized by antibiotic accumulation, release of free radicals, elimination of remaining metabolites, and irreversible loss of cell integrity and viability. Finally, the leaf senescence process is visible with necrosis of leaf cells or discoloration caused by chlorophyll degradation and is concluded with the death of the leaf cells (Dangl et al. 2000).

The leaf senescence can also be a defense response to pathogen infection. In this case, the plant reacts to the infection, inducing the death of leaf cells that are located around the infection isolating the pathogen and its damage. However, there are other forms of leaf senescence that can be induced by abiotic stresses, such as mechanical damage (wounds) or harvest.

Among the leaf senescence inducers the ethylene plays an important role. Ethylene is a simple gaseous plant hormone that regulates many diverse plant processes, during the whole plant development stages. The physiological ethylene effectiveness depends from environmental and biological factors, which may affect positively or negatively on the ethylene action.

3.2 Ethylene during Leaf Development

During the first leaf development stages, ethylene production is high and declines when leaves reach the fully expanded stage, and finally increases again during senescence (Fig. 3.1). The 1-aminocyclopropane-1-carboxylic acid (ACC) content, the precursor of ethylene, increases only in senescing leaves and has the same trend of ethylene evolution (Hunter et al. 1999). Analogous results were found measuring ethylene production from cut immature eucalyptus branches and mature branches (Ferrante et al. 1998). The amount of ethylene produced from cut immature branches (leaves not completely expanded) is about 3 nL g^{-1} h^{-1} while in mature branches (leaves completely expanded) is double. The cut branches placed in a vase show an increase of ethylene production during senescence until they reach 12 nL g^{-1} h^{-1} (Ferrante et al. 2003). Molecular approaches allowed isolating the genes that encode for ACC oxidase in white clover that catalyses the last step of ethylene biosynthesis. Three ACC oxidase genes were isolated and their expression was different during leaf development (Hunter et al. 1999).

However, ethylene produced by younger leaves does not induce any deleterious effect on the plant, suggesting that plants become sensitive when they reach the mature or senescence stages. Further molecular studies are needed in order to understand the relationship among ethylene receptors biosynthesis, ethylene production, and plant sensitivity during leaf development.

3.3 Ethylene-Induced Leaf Senescence

The first evidence of ethylene as a promoter of leaf senescence was observed in leaves and stems treated with ethylene. In particular, leaves exposed to ethylene may show yellowing, necrosis (death), and shattering. The first onsets of ethylene-induced senescence in leaves are the photosynthesis and chlorophyll reduction (Baardseth and von Elbe 1989). However, the ethylene response is variable and depends on the species, because every plant or part of a plant has different sensitivity to this hormone.

3.3.1 Leaf Yellowing

The visible symptom of leaf senescence is the loss of green color that shifts to yellow. This phenomenon is caused by chlorophyll degradation that is catalyzed by the chlorophyllase that converts the chlorophyll *a* and *b* to chlorophyllide and phytol (Matile et al. 1997).

Exogenous application of ethylene accelerates chlorophyll degradation in both attached and detached leaves in many cut flowers. Chrysanthemum flowers, for example, can be classified as sensitive and insensitive to leaf yellowing in relation to their response to exogenous ethylene. Cut chrysanthemum treated with 100 µL L^{-1} ethylene shows strong chlorophyll loss in the sensitive species, while no significant changes were observed in leaf color of the insensitive chrysanthemum (Reyes-Arribas et al. 2000; Ferrante et al. 2003). The dramatic effect of ethylene on chlorophyll loss has also been found in cut alstroemeria flowers exposed to exogenous ethylene (Ferrante et al. 2002a).

3.3.2 Leaf Abscission

The abscission of leaves is a highly coordinated phenomenon involving multiple changes in cell structure, metabolism, and gene expression. Abscission is referred to as the process of natural separation of leaf from the parent plant (Taylor and Whitelaw 2001). The first works on leaf abscission were focused on understanding plant hormones action in plants. The identification of ethylene as a plant hormone responsible for leaf drop came from the observation that plants relatively close to gas lamps lost their leaves (Abeles 1973). The ethylene mode action was studied in bean identifying the abscission zone and the anatomy characteristics.

Earlier studies discovered that abscission occurs in highly predictable zones involving cells that are morphologically distinct before the abscission event (Fig. 3.2).

The leaf abscission is influenced by auxin and ethylene evolution (Osborne 1955; Abeles and Rubinstein 1964; Morgan and Hall 1964). Ethylene controls leaf abscission but works in relationship with leaf auxin content.

The antagonistic relationship between ethylene and auxin is evident in the abscission zone where the auxin level seems to control the cell sensitivity to ethylene. Therefore, the balance between auxin and ethylene is a crucial point for leaf abscission (Fig. 3.3). During senescence, the leaves produce less auxin

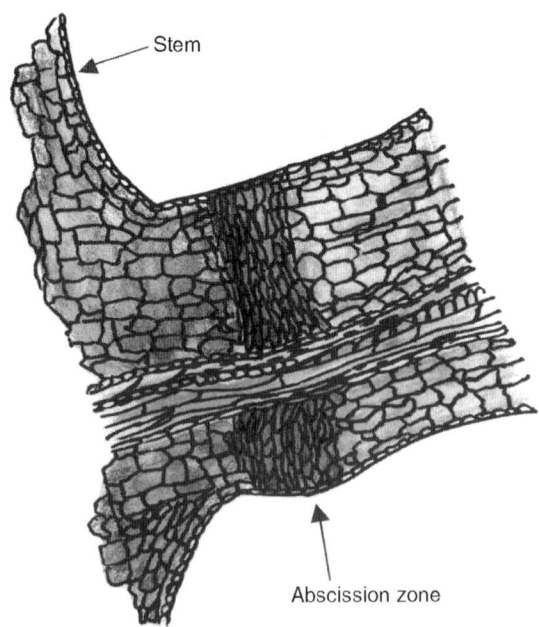

Fig. 3.2. Abscission zone between pedicel and stem

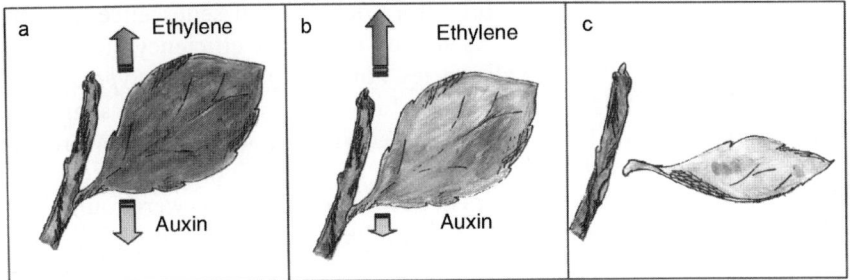

Fig. 3.3. Progressive leaf senescence. **a** leaf at fully expanded stage characterized by high auxin levels in equilibrium with ethylene production. **b** Initiation of leaf senescence, auxin content decreases and ethylene production increases. Leaf cells in the abscission zone become sensitive to ethylene action. **c** Leaf abscission

and make the abscission zone more sensitive to ethylene. Moreover, ethylene is an inhibitor of auxin biosynthesis and its increase during senescence accelerates the process. Application of exogenous auxin is possible to reduce leaf abscission as it was demonstrated in holly (Joyce et al. 1990). The abscission is a phenomenon where the enzymes produced by the cells digest the cellulose and other components of the cell walls. Many studies were performed to examine the cell wall-degrading enzymes that might be involved in cell wall degradation with evidence for the involvement of hydrolytic enzymes such as pectinases and cellulase.

3.4 Effect of Ethylene Inhibitors, Promoters, and Ethylene-Releasing Compounds on Leaf Senescence

Ethylene inhibitors reduce or eliminate the biological activity of ethylene. These compounds can be divided into two groups: inhibitors of ethylene biosynthesis and inhibitors of ethylene action. The first are substances that interact with the ethylene biosynthesis pathway through inhibition of key enzymes, ACC synthase, and ACC oxidase. The 1-aminoethoxyvinylglycine (AVG) and the amino-oxyacetic acid (AOA) are inhibitors of ethylene biosynthesis, while silver thiosulfate (STS), silver nitrate, and 1-methylcyclopropene (1-MCP) are inhibitors of ethylene action.

STS was first used for preventing ethylene action in cut flowers 30 years ago (Veen and van de Geijn 1978). The use of STS in agriculture is under government restriction, and its use, in some countries, is limited to potted plants. Silver is a heavy metal that persists in soil and may be a hazardous water pollutant as well as damage the nervous system in humans at high concentrations.

The application of ethylene inhibitors delayed leaf senescence in many ornamental species (Table 3.1). As already noted, the efficiency of the treatments may not be the same in all species. For example, cut *Eucalyptus gunnii* branches treated with 2 mM STS showed no increase in vase life, suggesting that this species has a low sensitivity to ethylene (Forrest 1991). Moreover, STS was not effective for cut evergreens and even induced phytotoxicity (Tingley and Price 1990). In contrast, STS pulse treatments were found to be beneficial for *Asparagus densiflorus, Cordyline terminalis,* and *Philodendron* "Red Emerald" (Broschat and Donselman 1987). Positive effects were also found in cut *Ilex aquifolium* and in *Phoradendron tomentosum*. This species treated with 0.2–2 µM STS dramatically reduced leaf fall (Joyce et al. 1990).

The last commercial ethylene action inhibitor is 1-methylcyclopropene (1-MCP), which has been extensively evaluated for protecting ornamentals from ethylene damage (Feng et al. 2004). This compound is a gas and is thought to bind the ethylene receptor protein, avoiding ethylene action (Fig. 3.4).

Treatments with 1-MCP have efficiently reduced the percentage of leaf abscission in citrus (Zhong et al. 2001) or inhibiting the leaf yellowing of cut lilies (Celikel et al. 2002).

Among the inhibitors of ethylene biosynthesis the most frequently used has been AOA, an inhibitor of ACC synthase that blocks the conversion of S-adenosyl methionine (SAM) to ACC. The AOA has been used for preventing

Table 3.1. Ethylene inhibitors used as treatment for extending vase life of cut flowers and foliage

Species	Compound	Concentration	References
Adiantum raddianum	Cobalt chloride	1 mM	Fujino and Reid (1983)
	AOA	0.2–0.5 mM	Fujino and Reid (1983)
Asparagus densiflorus	STS	2 mM	Broschat and Donselman (1987)
	AOA	0.01 mM	Dolci et al. (1989)
Cordyline terminalis	STS	2 mM	Broschat and Donselman (1987)
Eucalyptus gunnii	STS	2 mM	Forrest (1991)
Eucalyptus parvifolia	Cobalt chloride	2 mM	Ferrante et al. (2002b)
Gypsophila paniculata	MCP	200 ppb	Newman et al. (1998)
Ilex aquifolium	STS	12 mM	Joyce et al. (1990)
	NAA	33–66 ppm	Appleton et al. (1996)
Philodendron spp.	STS	2 mM	Broschat and Donselman (1987)
Phoradendron	STS	12 mM	Joyce et al. (1990)
Rumohra adiantiformis	Cobalt chloride	1 mM	Stamps and Nell (1986)

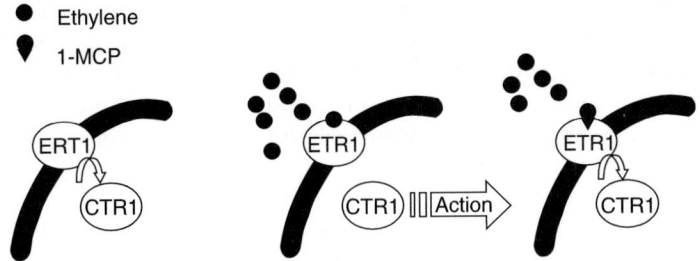

Fig. 3.4. Ethylene receptor and binding site on the cell membrane. The ethylene receptor (ETR1) contains the binding site for ethylene or 1-MCP. Ethylene gas binds the ETR1 and activates the constitutive triple response 1 (CTR1), inducing physiological responses (response negatively regulated). Plants treated with 200–500 ppb 1-MCP become ethylene insensitive, because the ethylene gas does not find binding sites available and does not induce any physiological effect

leaf senescence of many floriculture items such as cut *Asparagus plumosus* (Dolci et al. 1989). Cobalt nitrate is another inhibitor that has been also used for preserving cut *Adiantum raddianum* fronds (Fujino and Reid 1983). On the contrary, cobalt chloride has not affected the vase life of *Rumohra adiantiformis* fronds (Stamps and Nell 1986). These results show that each species has its own behavior and the same treatment may have a different effect depending on the species.

Ethylene promoters have been used in agriculture for several purposes. The most useful application is involved in olive fruits harvesting. Olives are non-climacteric fruits, and ethylene production by ripening olives has been reported to be non-detectable (Rugini et al. 1982). In many tissues that produce very little ethylene, the production of this hormone can be increased by exogenous application of ACC, suggesting that these tissues have the capacity to produce ethylene but do not do so due to a limited production of ACC (Kende 1993). Ethylene-releasing chemicals, used to promote olive fruit abscission, have been found to increase the efficiency of mechanical harvesting (Denney and Martin 1994; Gerasopoulos et al. 1999). However, these compounds must be used in appropriate concentrations otherwise they may cause excessive leaf abscission, which can compromise the following year's production if more than 15–20% of the leaves are lost.

The earlier ethylene-releasing compounds were Omaflora and calcium carbide. In horticulture, they were used for promoting pineapple flowering. The most common ethylene-realizing compound widely used in agriculture is ethephon. It was discovered by Russian researchers and its effect on plant physiology was found later by Amchem Products and rapidly became a commercial product. It is chemically called 2-chloroalkylphosphonic acid ($C_2H_6ClO_3P$) and in solution at pH 4–5 or higher is hydrolyzed into chloride ion, phosphate and ethylene. In agriculture, ethephon is considered as "liquid ethylene" and has many practical applications such as to promote pre-harvest ripening top fruit, soft fruit, tomatoes, and sugar beet. Extensive use of ethephon is localized in tropical regions for coffee, pineapple, rubber, and sugarcane production (Abeles et al. 1992).

Ethylene-releasing compounds are used for promoting fruit ripening such as tomato and bananas, or flowering such as pineapple, *Bromeliaceae*.

3.5 Leaf Sensitivity and Ethylene Production

Molecular studies performed on *Arabidopsis* have demonstrated that ethylene perception in plants is mediated by a family of *et*hylene *r*eceptors, including the ETR1, ETR2, (*e*thylene *r*esponse *s*ensor) ERS1, ERS2, and (*e*thylene *in*sensitive) EIN4 gene products (Hua and Meyerowitz 1998). These proteins located on the cell membrane are typically composed of a sensor protein and a response regulator protein, which function together to regulate adaptive responses to a broad range of environmental signals (Sakakibara et al. 2000).

The effects of ethylene can be observed when the amount of ethylene is higher then the plant sensitivity threshold. Most plants are considered ethylene sensitive if they show leaf senescence when exposed to 0.5–1 µL L^{-1} ethylene. Ethylene effects have a similar dose-response curve: no effect between 0.001 to 0.01 µL L^{-1} discernible effects between 0.01 and 0.1 µL L^{-1}, half-maximal responses between 0.1 to 1 µL L^{-1}, and saturation from 1 to 10 µL L^{-1} (Abeles et al. 1992).

The sensitivity of ornamental plants to exogenous ethylene was investigated by exposing them to 0–15 µL L^{-1} ethylene for 24 or 72 h in darkness at 20 °C (Woltering 1985, 1987; Woltering and van Doorn 1988). Results obtained showed that flowering plants were generally more sensitive to ethylene treatment than foliage plants. However, the ethylene effect was visible as abscission of flowers, flower buds, or whole inflorescence after 24 h. Foliage plants showed abscission and yellowing of leaves after 72 h. Ethylene toxicity symptoms have been described and 52 species are classified according to their ethylene sensitivity (Tingley and Price 1990).

The sensitivity threshold for each plant can be determined by using 10 µL L^{-1} exogenous ethylene; higher concentrations are not found in nature or closed environments, therefore plants that are sensitive to concentrations higher than that can be considered insensitive to ethylene.

The ethylene production varies from species to species and can be used for classifying plants in different groups depending on their levels of ethylene production. Evergreen species have been subdivided into six groups; the higher values were found in *Sequoia sempervirens* (2,800 nL $kg^{-1} h^{-1}$) and the lower in *Juniperus virginiana* (26 nL $kg^{-1} h^{-1}$).

However, on the same plant, the different organs have different sensitivity, therefore, leaves, flowers or other plant organs can be classified into four groups, considering their sensitivity to ethylene and their endogenous production:

- Sensitive plants that produce a high amount of ethylene;
- Sensitive plants that produce a low amount of ethylene;
- Insensitive plants that produce a low amount of ethylene;
- Insensitive plants that produce a high amount of ethylene.

Plants of *Sequoia sempervirens*, for example, are classified as high ethylene producers and also have low sensitivity to exogenous ethylene. On the contrary, cut *Ilex aquifolium* branches produce a very low amount of ethylene, ranging from 0.5 to 5 nL kg^{-1} h^{-1} during the first 96 h after harvest, and later the ethylene produced becomes undetectable (Joyce et al. 1990; Fjeld et al. 1995). Nevertheless, cut holly exposed to exogenous ethylene at concentrations as low as 0.001–0.6 µL L^{-1} for 2–3 days, completely loses its leaves.

3.6 Effect of Ethylene on the Antioxidant System during Leaf Senescence

The oxidative metabolism has an important role at cellular level when senescence takes place. During leaf senescence there is an overproduction of reactive oxygen species (ROS) such as superoxide anion ($O_2^{·-}$), hydroxyl radicals (OH$^·$), hydrogen peroxide (H_2O_2) and singlet oxygen ($^1O_2$) that may cause damage and cell death. However, they are also products of the normal enzymatic reactions in peroxisomes, glyoxysomes, and chloroplasts. The harmful ROS are controlled and balanced by the antioxidant systems present in leaves. The leaf cells act as a defense response against the accumulation of ROS by increasing catalase (CAT) and superoxide dismutases (SOD) activity. But other enzymatic and non-enzymatic antioxidants are also involved, such as ascorbate, reduced gluthatione (GSH), α-tocopherol and carotenoids (Foyer et al. 1994; Bartoli et al. 1996; Hodges et al. 1996, 1997a, 1997b). The ascorbate-glutathione cycle, also known as the Asada-Halliwell cycle, is the most important antioxidant system in the leaf cells during senescence (Fig. 3.5).

During leaf senescence, the enzymatic and non-enzymatic antioxidants are activated for reducing the ROS species that progressively increase during senescence (Kunert and Ederer 1985; Leshem 1988). In some herbaceous species, general leaf reductant levels were negatively correlated with the rate of senescence (Philosoph-Hadas et al. 1994; Meir et al. 1995). However, it has been found that some antioxidants increase while others decrease, depending on the species and the causes of leaf senescence (Irigoyen et al. 1992; Olsson 1995; Kingston-Smith et al. 1997).

It has been found that during natural senescence of chestnut leaves xanthine oxidase (XOD) and SOD increase significantly, while CAT and POD decrease progressively. Analogous studies carried out on flag leaf of different rice cultivars showed that CAT, SOD, and POD decreased in all cultivars (Jiao et al. 2003). These results suggest that during natural leaf senescence

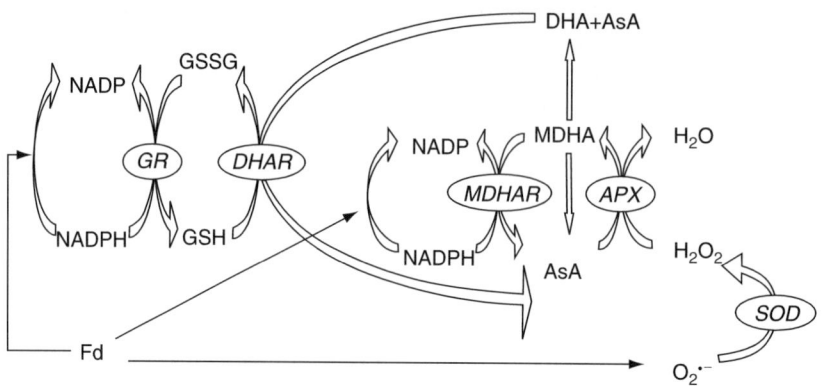

Fig. 3.5. Ascorbate-glutathione redox cycle in leaf senescence. The superoxide radicals ($O_2^{\cdot-}$) are dismutated by superoxide dismutase (SOD) that produces hydrogen peroxidase (H_2O_2). Ascorbate (AsA) is regenerated from oxidized forms by a cycle catalyzed by ascorbate peroxidase (APX). The AsA is converted in monodehydroascorbate (MDHA) by APX, then the monodehyroascorbate reductase (MDHAR) oxides the MDHA and reduces the $NADP^+$ that is regenerated by Ferredoxin (Fd). However, the MDHA may be converted to ASA and dehydroascorbate (DHA) by nonenzymatic reaction. The DHA is converted by dehydroascorbate reductase (DHAR) in ASA, contemporary the reduced glutathione (GSH) is transformed in oxidized glutathione (GSSG) by NADPH. During leaf senescence the NADPH production decreases, resulting in increased ratios of GSSG to GSH and DHA to AsA, causing changes in cellular redox conditions that may activate more SAGs

some antioxidant enzymes do not work in purpose because the leaves must follow their fate. On the contrary, these enzymes often increase when the senescence in younger leaves is induced by external factors (Monk et al. 1989; Foyer et al. 2002). In fact, during the first stages of leaf development, immediately after germination, SOD and POD activities are strongly induced when young wheat leaves are grown under stress conditions such as salinity (Khan 2003). These results confirm that before leaf senescence occurs, the antioxidant enzymes are activated for avoiding leaf damage, because plants need photosynthesis for their growth.

Ethylene may negatively affect the Asada-Halliwell cycle enzymes and accelerates leaf senescence (Bartoli et al. 1996). It has been found that 10 μL L^{-1} of ethylene applied to spinach leaves during storage significantly reduces ascorbate peroxidase (APX), CAT, dehydroascorbate reductase (DHAR), gluthatione reductase (GR), and monodehydroascorbate (MDHAR) activity (Burgheimer et al. 1967; Hodges and Forney 2000). Therefore, ethylene avoids the detoxification process and makes the senescence process more dramatic, leading the cells to death. However, the role of ethylene on the antioxidant cycle is not yet completely clarified. More experiments need to be carried out on mature leaves that are still able to photosynthesize and on senescent leaves. In mature leaves, photosynthetic electrons transfer generates NADPH, which donates electrons to AsA through GSH. In senescent leaves that do not photosynthesise the lack of ascorbate-glutathione cycle enzymes response might be due to NADPH decrease rather than to ethylene action.

3.7 Ethylene and Other Plant Hormones during Leaf Senescence

Ethylene can be naturally produced by any part of a plant, but can also be stimulated from other plant hormones such as auxins, gibberellins, abscisic acid, and cytokinins. In some tissues, the applications of exogenous synthetic plant hormones may increase ethylene evolutions and have a synergistic effect with ethylene. Ethylene production is predominantly enhanced by exogenous application of high concentrations of auxins (Yu and Yang 1979; Woeste et al. 1999; Vandenbussche et al. 2003). In *Phaseolus vulgaris*, the level of auxin in senescent leaves is ten times that of younger leaves (Atsumi and Hayashi 1979) and increases ethylene evolution.

Cytokinins and gibberellins are also substances that when applied exogenously increase ethylene production. Cytokinins or substituted of phenyl-urea compounds with cytokinin-like activity strongly increase ethylene production. Among the substituted of phenyl-urea compounds the most important is the thidiazuron that is used as plant growth regulator. In the particular the ability of thidiazuron to stimulate ethylene production in plants is used for inducing cotton defoliation and facilitate the harvesting operations (Suttle 1986). Gibberellins also increase ethylene production, but not much evidence is reported in literature regarding leaf senescence.

Ethylene production as a response to exogenous applications of synthetic plant hormones is probably a reaction to the alteration of the natural equilibrium of endogenous hormones or might be a response to stress induced by the treatments.

3.8 Ethylene and Gene Expression during Leaf Senescence

The role of ethylene during leaf senescence was investigated physiologically using ethylene inhibitors, genetically using several ethylene-insensitive mutants such as ETR1, ER, EIN2, and EIN3 or transgenic plants that do not express the key enzymes of ethylene biosynthesis. All these approaches showed that blocking ethylene biosynthesis or action could delay leaf senescence symptoms (Zacarias and Reid 1990; Grbic and Bleecker 1995; Chao et al. 1997; Oh et al. 1997).

The isolation of ethylene-resistant or ethylene-insensitive mutants has been very helpful in understanding ethylene function during leaf senescence. Molecular studies demonstrated that most of the genes affected by ethylene are also senescence associated genes (SAGs) that are enhanced, activated, or repressed during senescence (Table 3.2).

The genes involved during leaf senescence have been isolated using differential display, cDNA screening, and subtractive hybridization techniques. In *Arabidopsis thaliana*, many SAGs have been isolated and among them, see1, a sulfide dehydrogenase, is strongly induced by ethylene treatments. However, there are other SAGs, such as SAG 12, 13, 14, 20, ERD1, pSEN3, and pSEN4, that are also affected by ethylene (Park et al. 1998; Weaver et al. 1998).

Table 3.2. Genes affected by ethylene, ethylene promoters, or ethylene-releasing compounds during leaf senescence

Gene	Putative function	Putative role	Effect of ethylene
SPG31	Cysteine proteinase	Proteolysis and nitrogen remobilization	Up-regulated
pSEN3	Polyubiquitins	Protein turnover	Up-regulated
pSEN4	Endoxyloglucan transferase	Cell wall degradation	Up-regulated
CHLase1	Chlorophyllase	Chlorophyll degradation	Up-regulated
SARK	Senescence-associated receptor-like kinase	Signal transduction	Up-regulated
CAB	Light-harvesting chlorophyll a/b-binding protein	Photosynthesis	Down-regulated
ELIP	Early light-induced protein	Assembly or repair the photosynthetic apparatus	Up-regulated
Sen1	Sulfide dehydrogenase	Cellular redox reaction	Up-regulated
GR-S-transferase	Glutathione-S-transferase	Detoxification	Up-regulated at early senescence stage
LLS1	–	Unknown	Up-regulated at latest senescence stage
–	Putative protein	Signal transduction	Up-regulated at latest senescence stage
–	NADH-ferrihemoprotein reductase	Detoxification	Up-regulated at early senescence stage

Moreover, in *Arabidopsis* leaves spayed with ethylene-realizing compound (ethephon) the expression of pSEN3 and pSEN4 is enhanced; pSEN3 and pSEN4 encode for polyubiquitin and peptide related to endoxyloglucan transferase, respectively.

Ethylene also increases chlorophyll degradation, by increasing chlorophyllase (Chlase) gene expression. Since Chlase is confined in the chloroplast membrane (Matile et al. 1997), its activation is subordinated to membrane senescence. Membranes lose their integrity and allow the Chlase to come in contact with chlorophylls. Therefore, ethylene probably increases lipid degradation and destabilizes the cell membranes.

A bean (*Phaseolus vulgaris*) mutant that retains chlorophyll during ripening does not show any changes in color and is defined as a stay-green phenotype. The leaf senescence proceeds in both plants but no differences are visible in chlorophyll content both in attached and detached leaves of bean mutant. Exogenous ethylene applications induce leaf wilting and abscission in wild-type bean, while a slight loss of chlorophyll and carotenoids are observed in stay-green phenotype leaves. The Chlase activity decreases in both bean plants after 2 days of treatment with 10 μL L^{-1} ethylene (Fang et al. 1998). Analogous results have been observed in green flesh tomato, a mutant that also does not show leaf yellowing as a stay-green phenotype. Applying ethylene also increases ethylene chlorophyll breakdown in wild-type, while slight chlorophyll loss occurs in the mutant. However, in both plants chlorophyll a/b (CAB) binding protein and rubisco gene expression decrease during leaf senescence in both plants (Akhtar et al. 1999).

A senescence-associated receptor-like kinase (SARK) has been also isolated in bean plants using differential display technique. This gene is exclusively expressed during leaf senescence and its mRNA level increases before the chlorophyll loss and the chlorophyll a/b binding protein mRNA level decreases. The exogenous applications of ethylene strongly induce SARK transcript and accelerate leaf senescence of bean plants (Hajouj et al. 2000).

In tobacco (*Nicotiana tabacum* L. cv. SR1), the early light-induced protein (ELIP) was isolated using the cDNA-amplified restriction fragment polymorphism method. ELIP is a member of the CAB gene family and is implicated in the assembly or repair of the photosynthetic apparatus in plants under stress conditions. Treatments with ACC, a precursor of ethylene, enhanced the transcript levels of ELIP-TOB and its expression increases; before that chlorophyll loss could be detected (Binyamin et al. 2001). Molecular studies, as observed in the previous physiology experiments, confirm that ethylene strongly affects the leaf chlorophyll content, inducing many genes that lead to chlorophyll degradation.

In senescent leaves of sweet potato a cysteine proteinase named SPG31 has been isolated and its function seems to be involved in proteolysis and nitrogen remobilization during the leaf senescence process. Ethylene treatments applied to mature green leaves strongly induce the SPG31 transcripts (Chen et al. 2002).

In *Arabidopsis* plants, four genes were isolated that were up-regulated by treatment with 1 mM ACC. Two of these genes, NADH-ferrihemoprotein reductase and glutathione-S-transferase were enhanced by ACC at the early senescence stage, while the lls1 and a putative protein were expressed in the latest stage of leaf senescence (Gepstein et al. 2003).

The role of ethylene during leaf senescence has been widely demonstrated. Blocking ethylene in sensitive plants delays leaf senescence, but once senescence is initiated, the regulation of SAGs does not differ from wild-type and insensitive plants, suggesting that an age-dependent senescence program does not involve the ethylene-dependent pathway (Nam 1997). However, ethylene seems to regulate the timing of leaf senescence to respond to external factors (Grbic and Bleecker 1995). In conclusion, ethylene is surely a promoter of leaf senescence but may also be a response to other senescence inducers and may have a synergic effect with them, dramatically accelerating leaf senescence in many sensitive species.

References

Abeles FB (1973) Ethylene in plant biology. Academic, New York
Abeles FB, Rubinstein B (1964) Regulation of ethylene evolution and leaf abscission by auxin. Plant Physiol 39:963–969
Abeles FB, Morgan PW, Saltveit ME (1992) Ethylene in plant biology, 2nd edn. Academic, New York
Akhtar MS, Goldschmidt EE, John I, Rodoni S, Matile P, Grierson D (1999) Altered patterns of senescence and ripening in *gf*, stay green mutant of tomato (*Lycopersicon esculentum* Mill.). J Exp Bot 50:1115–1122
Appleton BL, Spivey AD, French SC (1996) Virginia cut holly production pruning, harvesting and marketing. Virg Coop Ext Publ 8:430–470
Atsumi S, Hayashi T (1979) Examination of the pronounced increase in auxin content of senescent leaves. Plant Cell Physiol 20:861–865
Baardseth P, Von Elbe JH (1989) Effects of ethylene, free fatty acid, and some enzyme systems on chlorophyll degradation. J Food Sci 54:1361–1363
Bartoli CG, Simontacchi M, Montaldi E, Puntarulo S (1996) Oxidative stress, antioxidant capacity and ethylene production during ageing of cut carnation (*Dianthus caryophyllus*) petals. J Exp Bot 297:595–601
Binyamin L, Falah M, Portnoy V, Soudri E, Gepstein S (2001) The early light-induced protein is also produced during leaf senescence. Planta 212:591–597
Broschat TK, Donselman H (1987) Potential of 57 species of tropical ornamental plants for cut foliage use. Hort Sci 22:911–913
Buchanan-Wollaston V (1997) The molecular biology of leaf senescence. J Exp Bot 48:181–199
Buchanan-Wollaston V, Earl S, Harrison E, Mathas E, Navabpour S, Page T, Pink D (2003) The molecular analysis of leaf senescence: a genomics approach. Plant Biotech J 1:3–22
Burgheimer F, McGill JN, Nelson AI, Stienberg MP (1967) Chemical changes in spinach stored in air and controlled atmosphere. Food Tech 21:109–111
Celikel FG, Dodge LL, Reid MS (2002) Efficacy of 1-MCP (1-methylciclopropene) and promalin for extending the post-harvest life of oriental lilies (*Lilium* x "Mona Lisa" and "Stargazer"). Sci Hort 93:149–155

Chao Q, Rothenberg M, Solano R, Roman G, Terzaghi W, Ecker JR (1997) Activation of the ethylene gas response pathway in *Arabidopsis* by the nuclear protein Ethyleneinsensitive3 and related proteins. Cell 89:1133–1144

Chen GH, Huang LT, Yap MN, Lee RH, Huang YJ, Cheng MC, Chen SCG (2002) Molecular characterization of a senescence-associated gene encoding cysteine proteinase and its gene expression during leaf senescence in sweet potato. Plant Cell Physiol 43:984–991

Dangl JL, Dietrich RA, Thomas H (2000) Senescence and programmed cell death. In: Buchanan B, Gruissem W, Jones R (eds) Biochemistry & molecular biology of plants. Am Soc Plant Physiol, Rockville, pp 1044–1101

Denney JO, Martin GC (1994) Ethephon tissue penetration and harvest effectiveness in olive as a function of solution pH, application time, and BA or NAA addition. J Am Soc Hort Sci 119:1185–1192

Dolci M, Deambrogio F, Accati E (1989) Post-harvest life of stems of *Asparagus plumosus* (Baker). Adv Hort Sci 3:47–50

Fang Z, Bouwkamp JC, Solomos T (1998) Chlorophyllase activities and chlorophyll degradation during leaf senescence in non-yellowing mutant and wild type of *Phaseolus vulgaris* L. J Exp Bot 49:503–510

Feng XQ, Apelbaum A, Sisler EC, Goren R (2004) Control of ethylene activity in various plant systems by structural analogues of 1-methylcyclopropene. Plant Growth Regul 42:29–38

Ferrante A, Mensuali-Sodi A, Serra G, Tognoni F (1998) Ethylene production and vase life in cut *Eucalyptus* spp. foliage. Ital Hort 5/6:57–60

Ferrante A, Hunter DA, Wesley PH, Reid MS (2002a) Thidiazuron: a potent inhibitor of leaf senescence in Alstroemeria. Postharvest Biol Technol 25:333–338

Ferrante A, Mensuali-Sodi A, Serra G, Tognoni F (2002b) Effects of ethylene and cytokinins on vase life of cut *Eucalyptus parvifolia* Cambage branches. Plant Growth Regul 38:119–125

Ferrante A, Mensuali-Sodi A, Serra G, Tognoni F (2003) Treatment with thidiazuron for preventing leaf yellowing in cut tulips, and chrysanthemum. Acta Hort 624:357–363

Fjeld T, Melberg, Hogetveit WR (1995) Ethylene sensitivity and ethylene production in English holly (*Ilex aquifolium* L.). Acta Hort 405:306–313

Forrest M (1991) Post-harvest treatment of cut foliage. Acta Hort 298:255–261

Foyer CH, Descourvières P, Kunert KJ (1994) Protection against oxygen radicals: an important defence mechanism studied in transgenic plants. Plant Cell Environ 17:507–523

Foyer CH, Vanacker H, Gomez LD, Harbinson J (2002) Regulation of photosynthesis and antioxidant metabolism in maize leaves at optimal and chilling temperatures: review. Plant Physiol Biochem 40:659–668

Fujino DW, Reid MS (1983) Factors affecting the vase life of fronds of maidenhair fern. Sci Hort 21:181–188

Gan S, Amasino RM (1995) Inhibition of leaf senescence by autoregulated production of cytokinin. Science 270:1986–1988

Gepstein S, Sabehi G, Carp MJ, Hajouj T, Nesher MFO, Yariv I, Dor O, Bassani M (2003) Large-scale identification of leaf senescence-associated genes. Plant J 36:629–642

Gerasopoulos D, Metzidakis I, Naoufel E (1999) Ethephon sprays affect harvest parameters of "Mastoides" olives. Acta Hort 474:223–226

Grbic V, Bleecker AB (1995) Ethylene regulates the timing of leaf senescence in *Arabidopsis*. Plant J 8:595–602

Hajouj T, Michelis R, Gepstein S (2000) Cloning and characterization of a receptor-like protein kinase gene associated with senescence. Plant Physiol 124:1305–1314

Hodges DM, Forney CF (2000) The effects of ethylene, depressed oxygen and elevated carbon dioxide on antioxidant profiles of senescing spinach leaves. J Exp Bot 51:645–655

Hodges DM, Andrews CJ, Johnson DA, Hamilton RI (1996) Antioxidant compound responses to chilling stress in differentially sensitive inbred maize lines. Physiol Plant 98:685–692

Hodges DM, Andrews CJ, Johnson DA, Hamilton RI (1997a) Antioxidant enzyme responses to chilling stress in differentially sensitive inbred maize lines. J Exp Bot 48:1105–1113

Hodges DM, Andrews CJ, Johnson DA, Hamilton RI (1997b) Antioxidant enzyme and compound responses to chilling stress and their combining abilities in differentially sensitive maize hybrids. Crop Sci 37:857–863

Hua J, Meyerowitz EM (1998) Ethylene responses are negatively regulated by a receptor gene family in *Arabidopsis* thaliana. Cell 94:261–271

Hunter DA, Yoo SD, Butcher SM, McManus MT (1999) Expression of 1-aminocyclopropane-1-carboxylate oxidase during leaf ontogeny in white clover. Plant Physiol 120:131–141

Irigoyen JJ, Emerich DW, Sánchez-Diaz M (1992) Alfalfa leaf senescence induced by drought stress: photosynthesis, hydrogen peroxide metabolism, lipid peroxidation and ethylene evolution. Physiol Plant 84:67–72

Jiao D, Benhua JI, Li X (2003) Characteristics of chlorophyll fluorescence and membrane lipid peroxidation during senescence of flag leaf in different cultivars of rice. Photosynthetica 41:33–41

Joyce DC, Reid MS, Evans RY (1990) Silver thiosulfate prevents ethylene-induced abscission in Holly and Mistletoe. Hort Sci 25:90–92

Khan NA (2003) NaCl-inhibited chlorophyll synthesis and associated changes in ethylene evolution and antioxidative enzyme activities in wheat. Biol Plant 47:437–440

Kingston-Smith AH, Thomas H, Foyer CH (1997) Chlorophyll a fluorescence, enzyme and antioxidant analyses provide evidence for the operation of alternative electron sinks during leaf senescence in a stay-green mutant of *Festuca pratensis*. Plant Cell Environ 20:1323–1337

Kende H (1993) Ethylene biosynthesis. Annu Rev Plant Physiol Plant Mol Biol 44:283–307

Kunert KJ, Ederer M (1985) Leaf aging and lipid peroxidation: the role of the antioxidants vitamin C and E. Physiol Plant 65:85–88

Leshem YY (1988) Plant senescence processes and free radicals. In: Pryor WA (ed) Free radical biology and medicine. Pergamon Press, New York, pp 39–49

Lutts S, Kinet JM, Bouharmont J (1996) NaCl-induced senescence in leaves of rice (*Oryza sativa* L.) cultivars differing in salinity resistance. Ann Bot 78:389–398

Matile P, Schellenberg M, Vicentini F (1997) Localization of chlorophyllase in the chloroplast envelope. Planta 201:96–99

Meir S, Kanner K, Akiri B, Philosoph-Hadas S (1995) Determination and involvement of aqueous reducing compounds in oxidative defense systems of various senescing leaves. J Agric Food Chem 43:1813–1819

Monk LS, Fagerstedt KV, Crawford MM (1989) Oxygen toxicity and superoxide dismutase as an antioxidant in physiological stress. Physiol Plant 76:456–459

Morgan PW, Hall DC (1964) Accelerated release of ethylene by cotton following application of indolyl-3-acetic acid. Nature 201:99

Nam HG (1997) The molecular genetic analysis of leaf senescence. Curr Opin Biotech 8:200–207

Newman JP, Dodge L, Reid MS (1998) Evaluation of ethylene inhibitors for postharvest treatment of *Gypsophila paniculata* L. Hort Tech 8:58–63

Noodén LD, Guiamét JJ (1989) Regulation of assimilation and senescence by the fruit in monocarpic plants. Physiol Plant 77:267–274

Oh SA, Park JH, Lee GI, Paek KH, Park SK, Nam HG (1997) Identification of three genetic loci controlling leaf senescence in *Arabidopsis thaliana*. Plant J 12:527–535

Olsson M (1995) Alterations in lipid composition, lipid peroxidation and anti-oxidative protection during senescence in drought stressed plants and non-drought stressed plants of *Pisum sativum*. Plant Physiol Biochem 33:547–553

Osborne DJ (1955) Acceleration of abscission by a factor produced in senescent leaves. Nature 176:1161–1163

Park JH, Oh SA, Kim YH, Woo HR, Nam HG (1998) Differential expression of senescence-associated mRNAs during leaf senescence induced by different senescence-inducing factors in *Arabidopsis*. Plant Mol Biol 37:445–454

Philosoph-Hadas S, Meir S, Akiri B, Kanner J (1994) Oxidative defense systems in leaves of three edible herb species in relation to their senescence rate. J Agric Food Chem 42:2376–2381

Reyes-Arribas T, Barrett JE, Huber DJ, Nell TA, Clark DG (2000) Leaf senescence in a non-yellowing cultivar of chrysanthemum (*Dendranthema grandiflora*). Physiol Plant 111:540–544

Rugini E, Bongi G, Fontanazza G (1982) Effects of ethephon on olive ripening. J Am Soc Hort Sci 107:835–838

Sakakibara H, Taniguchi M, Sugiyama T (2000) His-Asp phosphorelay signaling: a communication avenue between plants and their environment. Plant Mol Biol 42:273–278

Smart CM (1994) Gene expression during leaf senescence. New Phytol 126:419–448

Stamps RH, Nell TA (1986) Pre- and poststorage treatment of cut leatherleaf fern fronds with floral preservatives. Proc Fla State Hort Soc 99:260–263

Suttle JC (1986) Cytokinin-induced ethylene biosynthesis in nonsenescing cotton leaves. Plant Physiol 82:930–935

Taylor JE, Whitelaw CA (2001) Signals and abscission. New Phytol 151:323–339

Thompson JE, Froese CD, Madey E, Smith MD, Hong Y (1998) Lipid metabolism during plant senescence. Prog Lipid Res 37:119–141

Tingley DR, Price TA (1990) Ethylene production and influence of silver thiosulfate on ethylene sensitivity of cut evergreens. Hort Sci 25:944–946

Trippi V, Thimann KV (1983) The exudation of solutes during senescence of oat leaves. Physiol Plant 58:21–28

Vandenbussche F, Smalle J, Madeira-Saibo NJ, Paepe NJM, de Chaerle A, Tietz L, Smets O, Laarhoven R, Harren LJJ, Onckelen FJM, van Palme H, Verbelen J-P[Page No. 18], van der Straeten D (2003) The *Arabidopsis* mutant alh1 illustrates a cross talk between ethylene and auxin. Plant Physiol 131:1228–1238

Veen H, van de Geijn SC (1978) Mobility and ionic form of silver as related to longevity in cut carnation. Planta 140:93–96

Weaver LM, Gan S, Quirino B, Amasino RM (1998) A comparison of the expression patterns of several senescence-associated genes in response to stress and hormone treatment. Plant Mol Biol 37:455–469

Woeste KE, Vogel JP, Kieber JJ (1999) Factors regulating ethylene biosynthesis in etiolated *Arabidopsis thaliana* seedlings. Physiol Plant 105:478–480

Woltering EJ (1985) Sensitivity of various foliage and flowering potted plants to ethylene. Acta Hort 181:489–492

Woltering EJ (1987) Effects of ethylene on ornamental pot plants: a classification. Sci Hort 31:283–294

Woltering EJ, van Doorn WG (1988) Role of ethylene in senescence of petals: morphological and taxonomical relationships. J Exp Bot 39:1605–1616

Yoshida S (2003) Molecular regulation of leaf senescence. Curr Opin Plant Biol 6:79–84

Yu YB, Yang SF (1979) Auxin-induced ethylene production and its inhibition by aminoethoxyvinylglycine and cobalt ion. Plant Physiol 64:1074–1077

Zacarias L, Reid MS (1990) Role of growth regulators in the senescence of *Arabidopsis thaliana* leaves. Physiol Plant 80:549–554

Zhong GY, Huberman M, Feng XQ, Sisler EC Holland D, Goren R (2001) Effect of 1-methylciclopropene on ethylene-induced abscission in citrus. Physiol Plant 113:134–141

4 Effect of Ethylene on Adventitious Root Formation

JINXIANG WANG[1], RUICHI PAN[2]

4.1 Introduction

Ethylene is the simplest structure plant hormone, the biological role of which on plant development was discovered over a century ago. To date, it has been well documented that ethylene is versatile signaling molecule and plays an important role in many aspects such as apical hook formation, seed germination, leaf senescence, root growth and stress adaptations (Abeles et al. 1992).

The first step of ethylene synthesis is the conversion of methionine to S-adenosylmethionine (SAM), catalyzed by the SAM synthase; the conversion of SAM to 1-aminocyclopropane-1-carboxylic acid (ACC) is then catalyzed by the enzyme ACC synthase (ACS). Lastly, ACC is oxidized to ethylene by ACC oxidase (ACO) (Yang and Hoffman 1984; Kende 1993). All vascular plants tested to date synthesize ethylene via this pathway (Fig. 4.1). Recently much literature has indicated that the activity of ACS is highly regulated and closely parallels the level of ethylene biosynthesis (Chae et al. 2003; Wang et al. 2004).

Adventitious root formation is typically generated from phloem or innercortex parenchyma cells of epicotyls or hypocotyls of plant. So the formation of primordium is very important for adventitious rhizogenesis. Normally adventitious root formation can be divided into four different stages, namely induction phase, early initiation phase, late initiation phase, and growth and development phase (Jarvis 1986). Considerable research demonstrated that that ethylene is involved in the rhizogenesis since Zimmerman and Hitchcock (1933) for the first time found that ethylene stimulated adventitious root formation. However, the literature on the effects of ethylene on root formation is conflicting. Ethylene and ethylene-releasing compounds have been reported to promote (Roy et al. 1972; Robbins et al. 1983, 1985; Bollmark and Eliasson 1990; Pan et al. 2002), inhibit (Coleman et al. 1980; Geneve and Heuser 1983; Nordström and Eliasson 1984) or have no

---

[1] College of Resources and Environment and Root Biology Center, South China Agricultural University, Guangzhou, 510642 China

[2] College of Life Science, South China Normal University, Guangzhou 510631 China

Fig. 4.1. Ethylene Biosynthetic Pathway. AdoMet synthetase, ACCsynthase and ACC oxidase, respectively, catalyze each step shown with the *arrows*. AdoMet: S-adenyl-methionine; Met: methionine; ACC: 1-aminocyclopropane-1-carboxylic acid; MTA: methylthioadenine (from Schaller and Kieber 2002)

effect on root formation (Batten and Mullins 1978; Mudge and Swanson 1978; Harbage and Stimart 1996). The variable response to ethylene may have different reasons.

4.2 Ethylene Stimulates Adventitious Root Formation

Mung bean hypocotyl cuttings treated with indole-3-butyric acid (IBA) showed high levels of ACC, N-malonyl-ACC (MACC) and ethylene (Riov and Yang 1989). With the application of 2 mmol L^{-1} aminoethoxy vinyl glycine (AVG) to mung bean cuttings for 6 h, the formation of adventitious root and growth were inhibited. The number of adventitious root in *Picea abies* L. (Karst) hypocotyls cuttings was 64 when treated with 0.1 µmol L^{-1} ethephon (an ethylene-releasing compound) for 28 days, whereas 10 µmol L^{-1} $CoCl_2$ treatment inhibited ethylene and decreased the root number to 2 from 22 in the control. Endogenous ethylene has been found positively correlative with the root number (Bollmark and Eliasson 1990). Sun and Bassuk (1993) reported that IBA induced ethylene production and adventitious rooting in *Rosa hybrida* L. cuttings. Application of 10 mL L^{-1} ethylene stimulated adventitious root formation of dwarf petunia explant while 0.01–1 mL L^{-1} ethylene played little effect on the rooting process (Dimasi-Theriou et al. 1993). Similarly, in sunflower cuttings ethylene-releasing compound stimulated rooting, and inhibitors of ethylene synthesis and action inhibited adventitious rooting (Liu et al. 1990). Application of 5–10 µmol L^{-1} AVG to tomato

(*Lycopersicon esculeutum*) explant inhibited rooting; 100 μmol L^{-1} salicylic acid which inhibited ACC oxidation to ethylene, severely impeded adventitious rooting of lavender cuttings (Mensuali-Sodi et al. 1995). Study on *Rumex palustris* cutting exposed to 2.5 μL L^{-1} and 5 mL L^{-1} ethylene formed 25 and 44 roots, respectively, after 7 days, and control cuttings formed four and eight roots (Visser et al. 1996). Rooting time was short and the rooting rate was high when hybrid cultivar of peach and almond, GF677, were cultured in a tube sealed with rubber stopper, resulting in increased ethylene level, in contrast to the sealed with cotton stopper (Marino 1997). Diao et al. (1999) reported that stimulatory effect of ethephon on adventitious rooting in mung bean hypocotyls cutting was dependent on treatment period, treatments of ethephon during 0–12, 12–24, 24–36, and 36–48 h. The treatments stimulated rooting and the treatments at 0–12 h and 36–48 h were more effective than any other treatments (Figs. 4.2, 4.3). Pan et al. (2002) verified that ethylene production peaked three times during adventitious rooting process and the value of peak was decreased one by one. Lorbiecke and Sauter (1999) demonstrated that deepwater rice survived a long time under submerged conditions and adventitious rooting in it was stimulated by ethylene (Fig. 4.4). Application of ACC increased adventitious root formation in stem cuttings from ethylene-insensitive Never-ripe (NR) tomato mutant and wild-type plants, however, NR cuttings formed fewer adventitious roots than wild-type cuttings, indicating the stimulatory effect of ethylene on rooting (Fig. 4.5) (Clark et al. 1999).

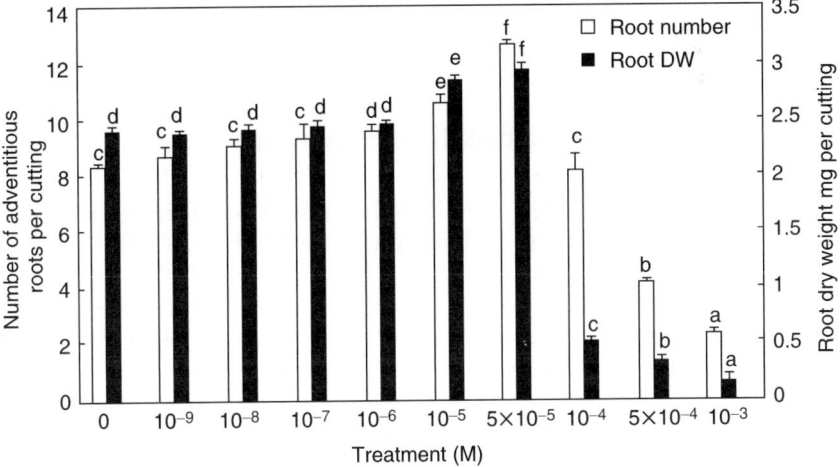

Fig. 4.2. Growth of adventitious roots at the third node of stem sections treated with ethephon as indicated. *Error bar* with different letter indicates significant difference (p<0.05) (from Pan et al. 2002)

Fig. 4.3. Adventitious roots at the third node of a deepwater rice shoot. Adventitious root initials are covered by the nodal epidermis during normal growth (*left*) or have emerged after treatment with 150 μM ethephon for 24 h (*right*) (from Mergemann and Sauter 2000)

4.3 Ethylene Inhibits Adventitious Rooting

Ethylene has also been shown to inhibit adventitious root formation in some plants. Rooting rate in cherry bud explant treated with 25, 50 and 250 μmol L^{-1} ACC under the presence of 5 μmol L^{-1} IBA was 58, 30 and 0%, respectively, and that of control was 78% (Biondi et al. 1990). Pea cutting treated with 0.1 μmol L^{-1} ACC and 10 μmol L^{-1} ACC for 4 days formed six and eight adventitious root, respectively, whereas the control formed 21 adventitious roots. Besides inhibition of root organogenesis in potato seedlings (Zacarias and Reid 1992) and grape cutting (Soulie 1993) has been reported by ethylene. Khalafalla and Hattori (2000) reported that $AgNO_3$ enhanced root emergence, root growth rate, root number per shoot, root length and improve rooting efficiency, while ACC reduced root number per shoot, root growth

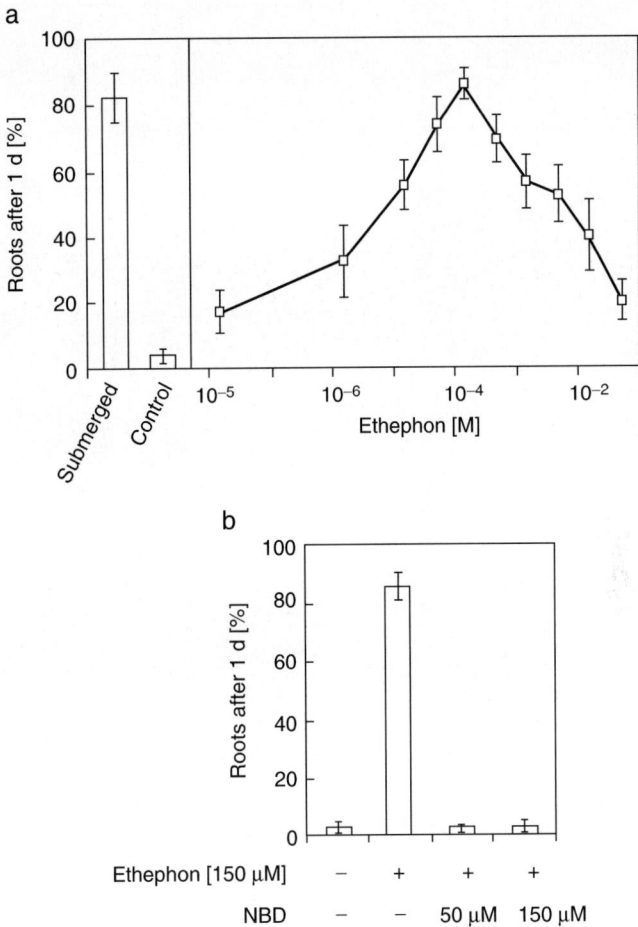

Fig. 4.4. Growth of adventitious roots at the third node of stem sections treated with ethephon as indicated for 24 h. **a** Effect of ethephon treatment at various concentrations compared with untreated sections (control) or with sections submerged for the same time (submerged). **b** Effect of ethephon treatment in combination with NBD, an inhibitor of ethylene action, at the concentrations indicated (from Lorbiecke and Sauter 1999)

rate, root length and consequently inhibited rooting efficiency. Since ethylene production in plant tissues increases following an application of ACC and $AgNO_3$ inhibits the action of ethylene, so they thought that ethylene inhibits root formation of faba bean shoots regenerated on medium supplemented with TDZ.

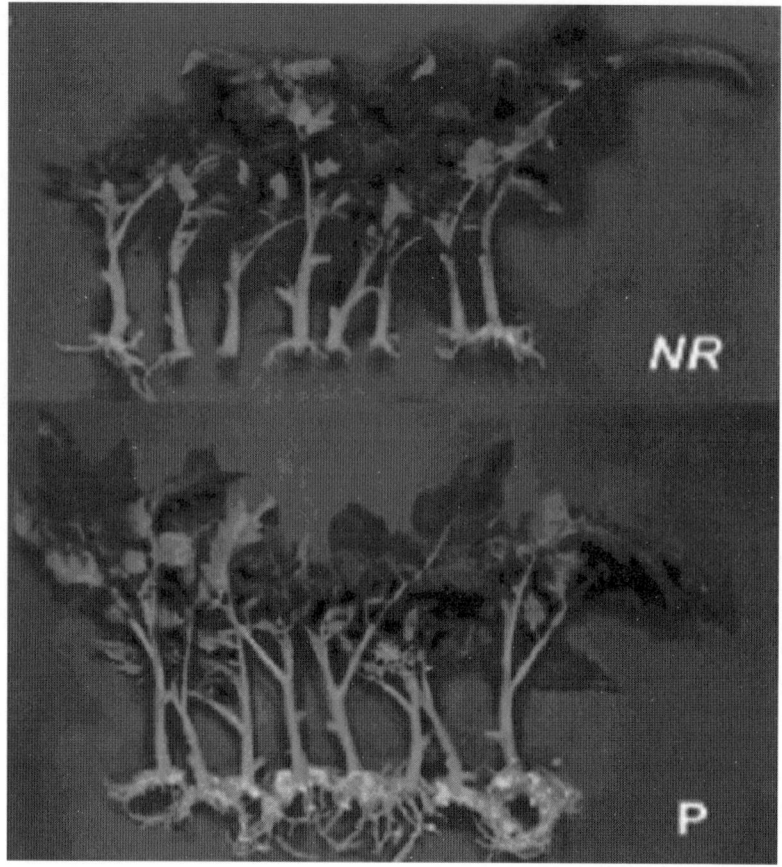

Fig. 4.5. Reduced adventitious root formation in potato mutant NR vegetative cuttings compared with wild-type cuttings. Stem cuttings were propagated for 21 days (from Clark et al. 1999)

4.4 Ethylene Has No Role in Adventitious Root Formation

Mudge and Swanson (1978) reported that application of ethephon (0–15 μL L^{-1}) to light-grown mung bean hypocotyl cuttings had no effect on adventitious rooting. The possible reason was that the concentration of ethylene was too low to stimulate rooting in mung bean cuttings.

4.5 Interactions between Ethylene and Other Hormones

Many kinds of plant growth regulators are involved in adventitious rooting. Ethylene plays a synergistic role with IBA in adventitious rooting of mung bean hypocotyl cuttings (Riov and Yang 1989). Synergism between plant growth retardants (inhibitor of gibberellin synthesis) and IBA in stimulating rooting of mung bean cuttings has been reported (Pan and Gui 1997). These results imply that interactions between plant hormones are pervasive.

Like ethylene, nitric acid (NO) is a versatile gaseous molecule that is involved in growth, pathogen defense, programmed cell death (PCD), and stress tolerance in higher plants (Pagnussat et al. 2004). Pagnussat et al. (2002) also reported that NO is involved in the auxin response during adventitious root organogenesis in cucumber and subsequently (Pagnussat et al. 2003) showed that an NO-mediated cGMP-dependent pathway was operating in the rooting process. Pagnussat et al. (2004) further verified that NO is upstream member of MAPK-mediated adventitious rooting in cucumber plant. Involvement of MAPK in ethylene signaling is well known (Schaller and Kieber 2002). It seems possible that NO is also involved in ethylene-induced adventitious rooting.

4.6 Discussion

How to interpret the contrasting effect of ethylene on adventitious rooting? We propose that the effect of ethylene on adventitious rooting is dependent on its concentration, time applied to cuttings, physiological conditions of treated cuttings, and endogenous ethylene level, and the interactions between other plant growth regulators. As hypothesized by Jusaitis (1986), there is an optimal ethylene concentration needed for rooting tissue. The ethylene concentration may be modulated by applying chemicals that alter the concentration.

The other possibility is that ethylene changes the rooting tissue sensitivity to auxin. Ethylene did not affect polar transport of auxin in decapitated sunflower (*Helianthus annus*) and did not slow auxin metabolism, resulting in unchanged auxin content, however, ethylene stimulated rooting of sunflower. These results indicated that ethylene changed rooting tissue sensitivity to auxin (Liu and Reid 1992). Visser et al. (1996) found that *Rumex palustris* Sm. produced more ethylene under submergence and increased rooting tissue sensitivity to auxin, thus accelerating initiation of adventitious roots. Rooting-stimulation effect of auxin was mediated by ethylene and tissue sensitivity to auxin was necessary to adventitious rooting. Another possibility may be that ethylene promotes rooting through stimulating cell cycle protein genes expression (Lorbiecke and Sauter 1999). The last may be that ethylene can change ratio of IAA/CTK. CTK content in Norway spruce *Picea abies* L(Karst)] grown under high light density was higher than that grown in low light density, ethylene increased breakdown of cytokinins (CTK) leading to an increase in rooting (Bollmark and Eliasson 1990) or ethylene-enhanced basic cellulase, thus leading to reducing mechanical resistance of the cortex and to the emergence of root initials (Linkins et al. 1973).

Ethephon at 150 µmol L^{-1} stimulated adventitious root growth significantly (Lorbiecke and Sauter 1999). Ethylene induced *cycB2:2* expression in the middle, root-containing tissue of the third node and there were differences between internode and node tissue related genes expression (Lorbiecke and Sauter 1999) suggesting that ethylene played different role between internode and adventitious root growth at node.

The apical meristem in adventitious root primordia was found activated before cell growth set in. A subpopulation of cells in G1 phase duplicated their genome in a synchronous manner between 4 and 6 h after submergence and then move to the G2 phase, entering mitosis probably around 10 h after submergence (Lorbiecke and Sauter 1999). Moreover, ethylene-induced gene activities were similar to the submergence-induced temporarily and spatially in deepwater rice. On the other hand, ethylene increased cellulase activity in some plant tissue (Campbell and Drew 1983; del Campillo et al. 1990; He et al. 1994). Ethylene might promote cell death and adventitious root growth by modulating some genes expression associated with cell cycle and probably enhanced activities of some enzymes related to degrade the cell wall and or membrane in the node of deepwater rice sections. Starch is the main storage product in rice as in other cereals and has to be degraded in order to provide an adequate carbohydrate and energy supply (Sauter 2000). It is possible that ethylene stimulates the breakdown of starch by down-regulation of ABA action in the deepwater rice stem sections providing energy and substance for forming new cells and adventitious root growth.

Suge (1985) founded that GA$_3$ played a synergistic role with ethylene in rooting. Interestingly, exogenously applied GA$_3$ up-regulated and ABA down-regulated *OsACS5* expression in both lowland and deepwater rice seedlings (van der Straeten et al. 2001). Within 1 h of submergence, the ABA level in petioles of *Rumex palustris* was reduced to 80% and this response was replaced by applied ethylene (Voesenek et al. 2003). For internode growth, exogenous ABA strongly inhibited growth, however, GA$_3$ reversed ABA effect (Hoffmann-Benning and Kende 1992). There are at least two possibilities for ABA to inhibit adventitious root growth and cell death. The first is via hindering the perception of exogenous ethylene, and the second through deterring the ethylene synthesis. Moreover, exogenous ABA reduced bromodeoxyuridine incorporation and mitotic events in root meristems of *Arabidopsis* and sunflower (Robertson et al. 1990; Leung et al. 1994). Exogenous application of ABA up-regulated *ICK1* expression, which interacted with both CDKA and cycD3 and then inhibited the histone H1 kinase activity of the complex. It caused dramatic growth inhibition and decreased the total number of cells per plant (Wang et al. 2004). Exogenous ABA inhibits G1/S transition in synchronized BY-2 cells and had no effect on further cell cycle progression when applied during S-phase (Swiatek et al. 2002). ABA prevents the DNA replication, keeping the cells in the G1 stage when applied just before the G1/S transition. It is also possible for ABA to negatively modulate cell division at the third node of deepwater rice and accordingly counteracted ethylene and GA$_3$ effect.

Adventitious root growth is partially attributed to cell division activated by some growth-promoting hormones via the G1 phase to enter DNA synthesis at an increased rate. Ethylene is directly responsible for adventitious root growth induction and cell death, and regulated genes expression (Lorbiecke and Sauter 1999; Sauter 2000). However, we do not know how deepwater rice transduces and reacts to the ethylene signal at the level of molecule and biochemistry during adventitious rooting process or why cell death is only located at the site of adventitious emergence. Moreover, it is also unclear if a secondary messenger (such as nitric oxide, which is necessary for auxin to stimulate adventitious root organogenesis) is involved in ethylene-stimulated adventitious rooting.

4.7 Outlook

Although considerable research demonstrated ethylene-regulated adventitious rooting, the effect of ethylene on adventitious root formation at the molecular and biochemical level is unclear. Moreover, the mechanisms involved in the interactions between ethylene and other plant-growth regulators (especially auxin) are not known. Attempts to clarify the interactions and intersections between ethylene and other hormones in adventitious rooting will partially depend on the newly characterized mutants that are insensitive to these hormones. Based on DNA microarray and chemical genetics, many signaling components involved in ethylene signal transduction during an adventitious rooting process will be determined and thus facilitate new insights into the mechanisms involved in the root-regulated effect of ethylene on adventitious root formation.

References

Abeles FB, Morgan PW, Salveit MEJ (1992) Ethylene in plant biology, 2nd edn. Academic, San Diego

Batten DJ, Mullins MG (1978) Ethylene and adventitious root formation in hypocotyl segments of etiolated mung bean (*Vigna radiata* (L)Wilczek) seedlings. Planta 138:193–197

Biondi S, Diaz T, Iglesias I (1990) Polymaines and ethylene in relation to adventitious root formation in *Prumus avium* shoot cultures. Physiol Plant 78:474–483

Bollmark M, Eliasson L (1990) Ethylene accelerates the breakdown of cytokines and thereby stimulated rooting in Norway spruce hypocotyls cuttings. Physiol Plant 80:534–540

Campbell R, Drew MC (1983) Electron microscopy of gas space (arenchyma) formation in adventitious roots of *Zea mays* L. subjected to oxygen shortage. Planta 157:350–357

Chae HS, Faure F, Kieber JJ (2003) The *eto1*, *eto2*, and *eto3* mutations and cytokinin treatment increase ethylene biosynthesis in *Arabidopsis* by increasing the stability of ACS protein. Plant Cell 15:545–559

Clark DG, Gubrium EK, Barrett JE, Nell TA, Klee HJ (1999) Root formation in ethylene insensitive plants. Plant Physiol 121:53–60

Coleman WK, Huxter TJ, Reid DM (1980) Ethylene as an endogenous inhibitor of root regeneration in tomato leaf disc culture *in vitro*. Physiol Plant 48:519–525

Del Campillo E, Reid PD, Sexton R, Lewis LN (1990) Occurrence and localization of 9.5 cellulase in abscising and nonabscising tissue. Plant Cell 2:245–254

Diao J, Wen FD, Pan RC (1999) Stimulatory effect of ethephon on rooting of mung bean hypocotyls cuttings. J South China Normal Univ (Natural science edition) (Chinese with English abstract) 1:102–105

Dimasi-Theriou K, Esonomou SA, Sfakiotakis EM (1993) Promotion of petunia (*Petunia hybrida* L.) regeneration *in vitro* by ethylene. Plant Cell Tiss Org Cult 32:219–22

Geneve RL, Heuser CW (1983) The relationship between ethephon and auxin on adventitious root initiation in cuttings of *Vigna radiata*. J Am Soc Hort Sci 108:330–333

Harbage JF, Stimart DP (1996) Ethylene does not promote adventitious root initiation on apple microcuttings. J Am Soc Hor Sci 12:880–885

He CJ, Drew MC, Morgan PW (1994) Induction of enzymes associated with lysigenous aerenchyma formation in roots of *Zea mays* during hypoxia or nitrogen starvation. Plant Physiol 105:861–865

Hoffmann-Benning S, Kende H (1992) On the role of abscisic acid and gibberrellin in the regulation of growth in rice. Plant Physiol 99:1156–1161

Jarvis BC (1986) Endogenous control of adventitious rooting in non-woody cuttings [A]. In: Jackson MB (ed) New root formation in plants and cuttings[M]. Nijhoff Publ, Dordrecht, pp 191–222

Jusaitis M (1986) Rooting response of mung bean cuttings to 1-aminocyclopropane-1-carboxylic acid and inhibitors of ethylene biosynthesis. J Sci Hort 29:77–85

Kende H (1993) Ethylene biosynthesis. Annu Rev Plant Physiol Plant Mol Biol 44:283–307

Khalafalla MM, Hattori K (2000) Ethylene inhibitors enhance *in vitro* root formation on faba bean shoots regenerated on medium containing thidiazuron. Plant Growth Regul 32:59–63

Leung J, Bouvire-Durand M, Morris PC, Guerrier D, Chefdor F, Giraudat J (1994) *Arabidopsis* ABA response gene ABI1: features of a calcium-modulated protein phosphatase. Science 264:1448–1452

Linkins AE, Lewis LN, Palmer RL (1973) Hormonally induced changes in the stem and petiol anatomy and cellulase enzyme patterns in Phaseolus vulgaris L. Plant Physiol 52:544–560

Liu JH, Reid DM (1992) Auxin and ethylene-stimulated adventitious rooting in relation to tissue sensitivity to auxin and ethylene production in sunflower hypocotyls. J Exp Bot 43:1191–1198

Liu JH, Mukherjee I, Reid DM (1990) Adventitious rooting in hypocotyls of sunflower (*Helianthus annuus*) seedlings. III. The role of ethylene. Physiol Plant 78:268–276

Lorbiecke R, Sauter M (1999) Adventitious root growth and cell-cycle induction of deep water rice. Plant Physiol 119:21–29

Marino G (1997) The influence of ethylene *in vitro* rooting of GF677 (*Prunus persica* Xprunus *amygdalus*) hybrid peach rootstock. In vitro Cellular Develop Biol Plant 33:26–29

Mensuali-Sodi A, Panizza M, Tognoni F (1995) Endogenous ethylene requirement for adventitious root induction and growth in tomato cotyledons and lavandin microcuttings *in vitro*. Plant Growth Regul 17:205–212

Mergemann H, Sauter M (2000) Ethylene induces epidermal cell death at the site of adventitious root emergence in rice. Plant Physiol 124: 609–614

Mudge KW, Swanson BT (1978) Effect of ethephon, indolebutyric acid and treatment solution pH on rooting and on ethylene levels within mung bean cuttings. Plant Physiol 61:271–273

Nordström AC, Eliasson L (1984) Regulation of root formation by auxin-ethylene interaction in pea stem cuttings. Physiol Plant 61:298–302

Pagnussat GC, Simontacchi M, Puntarulo S, Lamattina L (2002) Nitric oxide is required for root organogenesis. Plant Physiol 129:954–956

Pagnussat GC, Lanteri ML, Lamattina L (2003) Nitric oxide and cyclic GMP are messengers in the IAA-induced adventitious rooting process. Plant Physiol 132:1241–1248

Pagnussat GC, Lanteri ML, Lombardo MC, Lamattina L (2004) Nitric oxide mediates the indole acetic acid induction activation of a mitogen-activated protein kinase cascade involved in adventitious root development. Plant Physiol 135:279–286

Pan RC, Gui HY (1997) Physiological basis of the synergistic effects of IBA and triadimefon on rooting of mung bean hypocotyls. Plant Growth Regul 22:7–11

Pan RC, Wang JX, Tian XS (2002) Influence of ethylene on adventitious root formation in mung bean cuttings. Plant Growth Regul 36:135–139

Riov J, Yang SF (1989) Ethylene and auxin-ethylene interaction in adventitious root formation in mung bean (*Vigna radiata*) cuttings. J Plant Growth Regul 8:131–141

Robbins JA, Kays SJ, Dirr MA (1983) Enhanced rooting of wounded mung bean cuttings by wounding and ethephon. J Am Soc Hort Sci 108:325–329

Robbins JA, Reid MS, Paul JL (1985) The effect of ethylene on adventitious root formation in mung bean (*Vigna radiata*) cuttings. J Plant Growth Regul 4:147–157

Roy BN, Basu RN, Bose TK (1972) Interaction of auxin with growth-retarding: inhibiting and ethylene-promoting chemicals in rooting of cuttings. Plant Cell Physiol 13:1123–1127

Robertson JM, Hubick KT, Yeung EC, Reid DM (1990) Developmental responses to drought and abscisic acid in sunflower roots .2. Mitotic activity. J Exp Bot 41:339–350

Sauter M (2000) Rice in deep water: "how to take heed against a sea of troubles". Naturwissenschaften 87:289–303

Schaller GE, Kieber JJ (2002) Ethylene: the *Arabidopsis* book. Am Soc Plant Biol

Soulie O (1993) Ethylene inhibits the morphogenesis of *Vitis vinifera* cutting cultured *in vitro*. In: Pechjc, Latche A, Balag (eds) Cellular and molecular aspects of the plant hormone ethylene. Kluwer Academic Publ, Dordrecht, pp 367–368

Suge H (1985) Ethylene and gibberellin: regulation of internodal elongation and nodal root development in floating rice. Plant Cell Physiol 26:607–614

Sun WQ, Bassuk N (1993) Auxin-induced ethylene synthesis during rooting and inhibition of budbreak of 'Royalty' rose cuttings. J Am Soc Hor Sci 118:638–643

Swiatek A, Lenjou M, Bockstaele DV, Inzé D, Onckelen HV (2002) Differential effect of jasmonic acid and abscisic acid on cell cycle progression in tobacco BY-2 cells. Plant Physiol 128:201–211

Van der Straeten D, Zhou ZY, Prinsen E, Van Onckelen HA, Van Montagu MC (2001) A comparative molecular-physiological study of submerged response in lowland and deepwater rice. Plant Physiol 125:955–968

Visser EJW, Blom CWPM, Voesenek LACJ (1996) Flooding-induced adventitious rooting in *Rumex*: morphology and development in an ecological perspective. Acta Bot Neerl 45:17–28

Voesenek LACJ, Benschop JJ, Bou J, Cox MCH, Groeneveld HW, Millenaar FF, Vreeburg RAM, Peeters AJM (2003) Interaction between plant hormones regulate submergence-induced shoot Elongation in the flooding-tolerate dicot *Rumex palu-stris*. Ann Bot 91:205–211

Wang H, Zhou Y, Gilmer S, Whitwill S Fowke LC (2004) Expression of the plant cyclin-pendent kinase inhibitor ICK1 affects cell division, plant growth and morphology. Plant J 24:613–623

Yang SF, Hoffman NE (1984) Ethylene biosynthesis and its regulation in higher-plants. Annu Rev Plant Physiol Plant Mol Biol 35:155–189

Zacarias L, Reid MS (1992) Inhibition of ethylene action prevents root penetration through compressed media in tomato (*Lycopersicon esculentum*) seedlings. Physiol Plant 86:301–307

Zimmerman PW, Hitchcock AE (1933) Initiation and stimulation of adventitious roots caused by unsaturated hydrocarbon gases. Contr Boyce Thomp Inst 5:351–369

5 Ethylene and Plant Responses to Abiotic Stress

UWE DRUEGE

5.1 Introduction

Exposure to physical, chemical or biological stresses is an intrinsic characteristic of plant life not only in the natural environment but also under more or less controlled conditions in horticultural and agricultural systems. Globalisation and concentration of plant production throughout the world, on the one hand, allows for optimizing of growth conditions, but, on the other, increasingly involves post-harvest stress during storage and transport of intermediate and final products (Çelikel and Reid 2002; Kadner and Druege 2004). According to Levitt (1980), the term stress is used for any external factor capable of inducing a potentially injurious strain in the plant. With regard to the condition of the plant, plant stress can be described as a state in which increasing demands made on a plant lead to an initial destabilization of functions, followed by adaptive responses towards normalization, which may even lead to over-compensation of functions and improved resistance (Larcher 1995). However, if the acute tolerance or adaptive capacity is overtaxed, permanent damage or death may result. Within a network of hormonal cross-talk (Gazzarrini and McCourt 2003), ethylene production and signalling are highly involved not only in stress-induced symptoms of disturbed growth, senescence or injury, but also in acclimation processes which aid plant performance and survival.

5.2 Abiotic Stress and Ethylene Biosynthesis

One of the most common and pronounced responses of plants to various environmental stresses is the enhanced production of ethylene. Ethylene production of plants was shown to be accelerated by exposure to various physico-chemical stresses (Hyodo 1991; Abeles et al. 1992; Morgan and Drew 1997), some of which are also involved in enhanced ethylene production in response to biological stresses such as infection with pathogens or infestation

---

Institute of Vegetable and Ornamental Crops Grossbeeren/Erfurt e.V., Department Plant Propagation, Kuehnhaeuser Str. 101, 99189 Erfurt-Kuehnhausen, Germany

with herbivores (Boller 1991; Arimura et al. 2002). The ethylene biosynthetic pathway in higher plants is well established (Pech et al. 2004) and illustrated in Fig. 5.1. It starts with the synthesis of the intermediate S-adenosylmethionine (SAM) from methionine (Met) through an ATP-dependent step catalysed by SAM synthetase (SAMS). In the next step, 1-aminocyclopropane-1-carboxyl acid (ACC) is generated from SAM by ACC synthase (ACS), and Met is regenerated within the Yang's cycle. In addition to its role in ethylene synthesis, SAM is the major methyl group donor in numerous trans-methylation reactions and, in an alternative way, can be converted to decarboxylated SAM (dSAM), which is catalysed by the enzyme S-adenosylmethionine decarboxylase (SAMDC) (Hu et al. 2005). dSAM is a precursor of polyamines (PA), which may constitute to stress tolerance of plants (see Sect. 5.4). The direct precursor of ethylene, ACC is highly mobile with in the cell and, at the whole plant level can be translocated basipetally in the phloem or acropetally through the xylem. The last step of ethylene biosynthesis from ACC is catalysed by ACC oxidase (ACO) and requires oxygen and CO_2 as a co-substrate and essential activator, respectively. The toxic by-product cyanide is usually withdrawn from the cell by incorporation into L-cystein to form β-cyanolalanine, which is catalysed by β-cyanolalanine synthase (CAS) as enhanced by ethylene, and is further metabolized. In an alternative way, ACC can be diverted from its route to ethylene by forming conjugate derivatives. The major conjugate N-malonylACC (MACC) is not converted back to ethylene under normal conditions, while 1-(γ-L-glutamylamino)ACC (GACC) represents at most a minor form.

Genes of the critical enzymes have been increasingly characterized and cloned during the last decade so that it was possible to follow the regulation of ethylene biosynthesis up to the transcriptional level by studying (and also manipulating) gene expression. It is now recognized that the enzymes ACS and ACO are encoded by multi-gene families, of which genes are regulated independently and in a tissue-specific manner in response to one or a specific set of several environmental effectors (Fig. 5.1). Studies further highlighted the complexity that depending on the particular stress factor, plant, tissue and developmental stage, the generation of ACC and/or its conversion into ethylene are more or less stimulated, while the generation of SAM from Met is rather involved in a secondary or supportive way (Morgan and Drew 1997; Pech et al. 2004). The following sections address the response of the ethylene biosynthetic pathway to particular stresses, focusing on the transcriptional and also considering post-transcriptional regulation of ACS and ACO.

5.2.1 Wounding and Mechanical Stress

Injury is the most heavy stress, at least at the cellular level, and happens not only in the natural environment but also is frequently involved in harvesting and post-harvest handling of agricultural products such as cuttings, cut flowers,

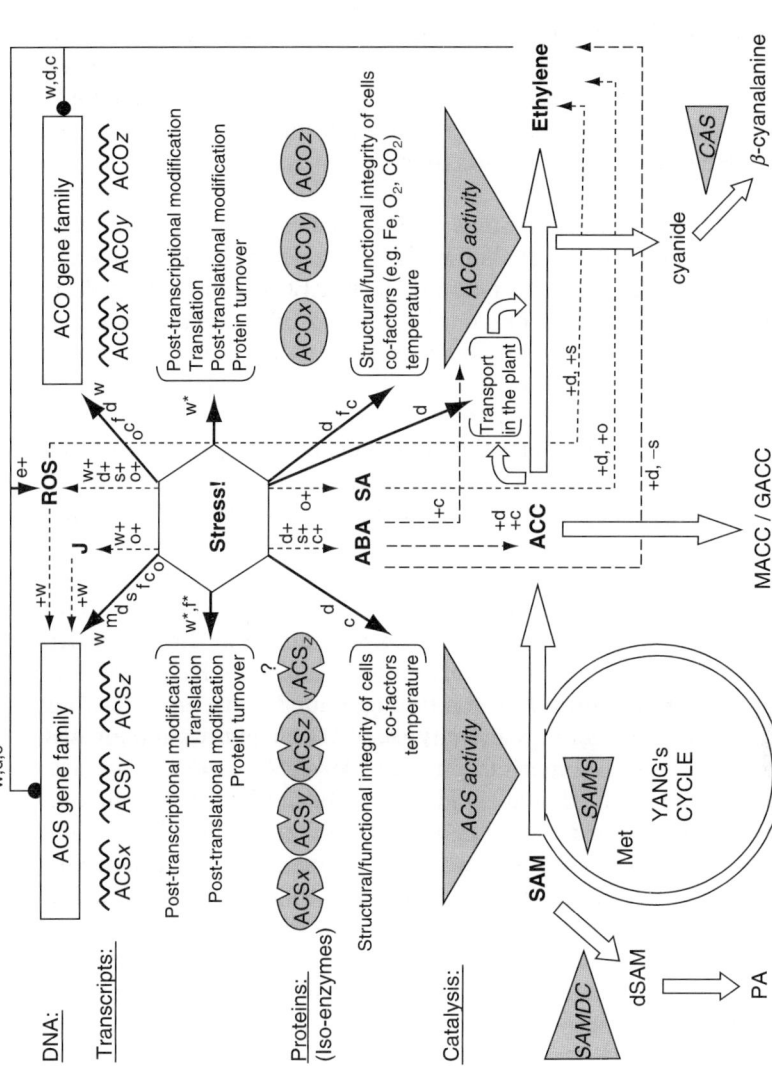

Fig. 5.1. A schematic model illustrating the ethylene biosynthetic pathway and the stress-mediated regulation of ACC and ethylene synthesis at gene, protein, cellular and plant level. *Arrowed lines* indicate the influence of stress. *Lines ending with a black circle* indicate a feedback control by ethylene. *Dashed arrowed lines* indicate the involvement of ROS, J, SA, and ABA as explained in the text. Examples of individual stresses (i) such as wounding (w), mechanic stress (m), water deficit (d), salt (s), flooding/hypoxia (f), chilling (c), and ozone (o) are included according to the text. +(−)i indicates a positive (negative) influence of the substance on the response to the individual stress i; i+ and e+ indicate that the stress i and ethylene, respectively, increased the concentration of the substance; * indicates that there is no direct evidence

plants, vegetables and fruits. Wounding of plant tissues for example by cutting, piercing, bruising, abrasion, breaking or bending greatly stimulates ethylene production and this response is nearly universal among higher plants. In addition, non-injury mechanical forces or impedance may be sufficient to accelerate ethylene biosynthesis in plant tissues. Mechanically stimulated ethylene production can cause secondary responses in experimental procedures (Morgan and Drew 1997). Injury-caused ethylene biosynthesis at the cellular level may also be involved in whole plant responses to other stresses.

5.2.1.1 Wound Ethylene in Fruits

Wound ethylene produced in fruits has been intensively studied because of its great relevance to post-harvest shelf life and ripening. This phenomenon has been utilized in plant production since long before knowing the working physiology behind it. Thus, the practice of gashing of sycamore fig (*Ficus sycomorus*) fruits over thousands of years in East-Mediterranean countries led to fruit ripening (Galil 1968), which later was associated with marked increase in ethylene production (Zeroni et al. 1972). Studies over the last decade showed that production of wound ethylene in fruits is controlled by a coordinated expression of both ACS and ACO genes, the latter was revealed to have often a positive feedback control by ethylene. Wounding of mature cucumber (*Cucumis sativus*) fruits by slicing and subsequent cutting induced the accumulation of mRNAs of the genes CS-ACS1, CS-ACS2 and CS-ACO1, while maximum expression of both CS-ACS1 and CS-ACO1 coincided with the peak of ethylene production (Shiomi et al. 1998). Cutting of the mesocarp tissue of pre-climacteric melon (*Cucumis melo*) fruits of two cultivars induced a dramatic production of wound ethylene, coinciding with increased ACC levels, ACS activity and ACO activity, and greatly induced mRNA accumulation of the genes CMe-ACS1 and CMe-ACO1, whereas CMe-ACO2 remained unaffected (Shiomi et al. 1999). Application of ethylene to the non-wounded fruits significantly induced CMe-ACO1 mRNA accumulation in both the cultivars, which supported a positive feedback control of wound-induced expression of ACO. Mathooko et al. (2001) studied the effect of combinations of wounding and applications of the translational inhibitor cycloheximide (CHX), the ethylene action inhibitor 1-methylcyclopropane (MCP) as well as ethylene on the ethylene biosynthetic pathway in peach (*Prunus persica*) fruits, and found an important role of primary, translation-independent, expression of genes of ACS and ACO in wound-induced ethylene production as well as a positive feedback control of wound-induced expression and activity of ACO. Stimulation of the wound-induced ACS activity and expression of PP-ACS1 after treatment with MCP indicated a negative feedback control of wound-induced ACS. Studies on avocadoes (*Persea americana*) also revealed a positive feedback of wound-induced expression of ACO and a negative feedback regulation of the

ACS gene PA-ACS2 by ethylene, which, however, was contrasted by a positive feedback regulation of the gene PA-ACS1 (Owino et al. 2002).

Kato et al. (2000) studied the limiting factors in wound-induced ethylene production in the mesocarp of squash (*Cucurbita maxima*) fruits, differentiating the layer with the wounded cut surface (layer 1, 1 mm thickness) from the same-sized, adjacent non-injured layer (layer 2) of halved fruits. In layer 1, maxima of ethylene production and ACC levels were detected at 16 h after wounding and were preceded by maximum ACS activity at 8 h, which followed the expression of CMa-ACS mRNA induced at 4 h. In layer 2, ethylene production reached only 25% of the value of layer 1. This was associated with an only marginal increase in the ACC level, almost constantly low ACS activity and only barely detectable expression of CMa-ACS. By contrast, ACO activity and expression of the gene CMa-ACO increased progressively in both of the layers. In layer 1, maximum values were reached at 24 h, whereas the values in layer 2 increased further to 36 h, strongly surpassing those of layer 1. Suppression of wound-stimulated induction of ACO activity and gene expression by application of the ethylene action inhibitor 2,5-norbornadiene (NBD) was most pronounced in layer 2 and indicated a positive feedback of ethylene. The results suggest that the induction of ACS and generation of ACC was of primary importance for the wound-induced ethylene biosynthesis (Kato et al. 2000).

Jasmonic acid (JA) and its methyl ester can induce ethylene production (O'Donnell et al. 1996) and the transcription of involved enzymes (Blume and Grierson 1997) in plant tissues, suggesting that a cross-talk occurs between the ethylene and jasmonate (J) signalling pathways. There is also increasing indication that a wide range of stress-induced disorders are mediated by reactive oxygen species (ROS) (Purvis et al. 1995). Watanabe and Sakai (1998) and Watanabe et al. (2001) provided evidence that both compounds are involved in the wound-induced ethylene production in squash (Fig. 5.1). They showed that ROS were produced immediately and transiently after wounding and that wound-induced expression of the ACS gene CM-ACS1 was stimulated by ROS and by methyl jasmonate (MJ). After wounding of the mesocarp issue, JA concentrations increased within 2 h, after 3 h reaching a 5-fold level compared to that of intact tissue. Application of JA to the wounded tissue promoted wound-induced ethylene biosynthesis and accumulation of CM-ACS1. When diphenyleneiodonium (DPI) and acetylsalicylic acid (AA) were used to inhibit either the superoxide-generating enzyme NAD(P)H oxidase or the synthesis of JA, respectively, application of both compounds to the wounded tissue reduced the wound-induced production of ethylene. However, DPI, which did not interfere with the JA accumulation, strongly inhibited the accumulation of the CM-ACS1 mRNA only during the early period of 1 and 2 h after wounding, whereas application of AA inhibited the mRNA accumulation only within 2 and 3 h after wounding. Watanabe et al. (2001) suggested from the studies that the wound-induced expression of the CM-ACS1 gene was initially mediated by the early increasing ROS levels and then by JA, the level of which increases at the later time after wounding.

5.2.1.2 Wound Ethylene in Vegetative Tissues

Particularly in vegetative tissues, such as leaves, production of wound ethylene often has an explosive but transient character leading to a burst of ethylene evolution within a few hours and a rapid and strong decline to low levels thereafter (Einset 1996; Shiu et al. 1998; Tatsuki and Mori 1999). Regulation of both ACS and ACO can be involved in the process (Fig. 5.1). However, there is only little information available about the coordination between both these events. Depending on the plant species, tissue, and even the particular experiment, wounding induced the transcription of only one member, a set of or even no (Yu et al. 1998; Jones and Woodson 1999) members of the studied respective gene families. Transcription of the ACS genes ACS1, ACS2, ACS4 and ACS5 in leaves of *Arabidopsis thaliana* was not responsive to cutting with scissors (Arteca and Arteca 1999). However, in the same study, a rapid and strong induction of the ACS6 gene was found, which was immediately followed by a substantial increase of ACC levels in same tissues. Transcription of ACS6 was also responsive to eight other factors including application of ethylene but also of aminooxyacetic acid (AOA), an inhibitor of ACC synthesis. In a recent study, transgenic lines of *Arabidopsis* were constructed expressing the β-glucuronidase (GUS) and green fluorescence protein (GFP) reporter genes from the promoter of each of the *Arabidopsis* ACS gene family members (Tsuchisaka and Theologis 2004). Reporter gene expression indicated that wounding of hypocotyls inhibited the constitutive transcription of the genes ACS1 and ACS5 in the same tissues, whereas expression of ACS2, ACS4, ACS6, ACS7, ACS8 and ACS11 was induced. Wang et al. (2005) focused on the expression of ACS4, ACS5 and ACS7 in the same plant species using ACS-GUS transgenic lines. They found that wounding of leaves by squeezing with tweezers caused an increased reporter gene activity only in the ACS4-GUS and the ACS5-GUS transformed plants. Since expression of ACS4 and ACS7 was stimulated by ethylene, the authors proposed a coexistence of ethylene autocatalysis and autoinhibition in vegetative tissues. The positive wound-response of ACS4-GUS activity contrasted to the missing wound response of ACS4 transcription in leaf tissue found by Arteca and Arteca (1999). This may be caused by the different detection methods but also by a different degree of injury. The influence of the latter is supported from results obtained with tomato (*Lycopersicon esculentum*). Wounding of leaflets with a wire brush caused a substantial increase of ethylene production, which was associated with increased expression of the genes LE-ACS7 and LE-ACS3, whereas LE-ACS2 did not respond (Shiu et al. 1998). In contrast, crushing of leaflets with a hemostat not only increased expression of LE-ACS6 and LE-ACS1A mRNA, the latter coinciding with maximum ethylene production, but also induced accumulation of LE-ACS2 (Tatsuki and Mori 1999). Also in *Nicotiana* species, wounding of leaves induced expression of a set of ACS genes in same tissues (Ge et al. 2000). The different genes revealed different kinetics falling into two classes of (i) rapid induction (peak after 2 h) combined with rapid disappearance

(NT-ACS4) and (ii) slow induction (peak after 6 h) combined with a longer persistence (NT-ACS2, NT-ACS3, NT-ACS5). It was suggested that there might exist two distinct wound-signaling pathways.

Wounding of sunflower (*Helianthus annuus*) seedlings by excision of roots increased ACO mRNA levels and ACO activity (Liu et al. 1997). Both events were consistent with enhanced ethylene production and were additionally stimulated by application of silver nitrate (SN) or silver-thiosulphate (STS) as ethylene action inhibitors. Applications of neither ACC nor the ethylene-releasing compound ethephon affected ACO mRNA levels, which did not support a positive feedback of wound-induced ACC or ethylene on ACO in these tissues. Gene expression studies on tomato revealed that transcription of ACO-genes in leaves increased in response to wounding, of which LE-ACO1 was the most abundantly expressed member (Barry et al. 1996). After transformation of tobacco and tomato fusing a GUS-reporter gene to an untranscribed LE-ACO1 sequence, reporter gene expression could be induced by wounding and most intense staining was restricted to the wound sites and to the cut surface of the detached leaf (Blume and Grierson 1997). Reporter gene expression was also enhanced by treatment with ethylene, which revealed a positive feedback of ethylene biosynthesis.

In leaves of melon, ACO activity and transcripts of the gene CM-ACO1 increased to a maximum at 2 h after wounding (Bouquin et al. 1997). Treatments with the translational inhibitor CHX or the ethylene action inhibitor MCP did not suppress the wound response of ACO transcription. However, apart from wounding, the activity and transcription of ACO could also be enhanced by ethylene, which was suppressed by MCP. Furthermore, fusing of differently sized fragments of the CM-ACO1 promoter to the GUS reporter gene in transgenic tobacco revealed two separate regions responsive to either ethylene treatment or wounding. The authors concluded that induction of CM-ACO1 gene expression in response to both factors occurs via two direct but separate signal-transduction pathways.

There is indication that the ACC generated by ACS may contribute to the wound-induced expression of ACO genes in leaves. Transcription of the ACO gene ACO1 in potato (*Solanum tuberosum*), which increased within 10 min after punching of the leaves with a metal brush was also induced by incubation of detached leaves in solutions of ACC, whereas the response of another gene, ACO2, was low upon both the treatments (Nie et al. 2002).

A recent paper of Katz et al. (2005) provided evidence that two types of ethylene production, an autocatalytic and an autoinhibitory system might operate in leaves, both of which could also be induced by wounding. After young, fully expanded leaves were detached from citrus plants (*Citrus sinensis*), a first phase of ethylene production up to 6 days after detachment was observed. This phase was characterized by a low and constant ethylene production associated with constitutive expression of the ACS gene CsACS2 and lacking expression of CsACS1, while autoinhibition of ethylene biosynthesis became evident from treatments with ethylene or propylene. The initial phase was followed by a climacteric rise in ethylene production associated with the induction of CsACS1 and CsACO1. This phase exhibited autocatalysis by increased transcription of both genes after application of ethylene or propylene. Both phases of ethylene production were also detected after additional wounding of leaves by excision and puncture of leaf discs that produced peaks after 15 min and 6 h, respectively. The principally similar response but of different kinetics of the two phases after detachment of young leaves, mature leaves and after strong local injury indicates that the developmental stage as well as the intensity of injury may be of great importance for the kinetic of wound-induced ethylene biosynthesis in leaf tissues.

There are indications that the wound-stimulated activity of the enzymes of the ethylene biosynthesis pathway is mediated by post-transcriptional/translational control and/or enzyme turnover (Fig. 5.1). Such processes are suggested by the findings that wound-induced increases in LE-ACO1-GUS-reporter gene expression in tomato (Blume and Grierson 1997), in ACO activity in melon leaves (Bouquin et al. 1997) and peach fruits (Mathooko et al. 2001), and in ACO protein levels in sunflower hypocotyls (Liu et al. 1997) were much lower or even absent when compared to the respective rises in mRNA accumulation. Tatsuki and Mori (2001) showed that LE-ACS2 protein, a wound-inducible isoenzyme of ACS in tomato fruit is phosphorylated at the C-terminal region in vivo. Considering that the enzymatic activity did not change after in vitro phosphorylation, the authors suggested that the phosphorylation of the C-terminal region might regulate the turnover of ACS isozymes.

5.2.1.3 Mechanical Non-Injury Stress

It has been repeatedly shown that mechanical, obviously non-injury, stimulation of plants, for example by touching, rubbing, shaking, brushing or wind can increase ethylene biosynthesis in plants (Hyodo 1991). Recent papers reveal that transcription of some ACS genes sensitive to wounding can also be induced by mechanical stimulation (Fig. 5.1). In *Arabidopsis thaliana*, touching of leaves by bending over a 30-s period not only increased endogenous ACC levels but also induced a transient accumulation of transcripts of the multi-responsive ACS gene ACS6, which showed a kinetic similar to that induced by wounding (Arteca and Arteca 1999). Also in tomato, mechanic stimulation by striking the leaves back and forth and rubbing the green fruits induced ethylene production, which followed a marked increase in the levels of LE-ACS6 and LE-ACS1A mRNA transcripts, whereas LE-ACS2 only responded to severe cell damage (Tatsuki and Mori 1999). Because the kinetics in response to touch was very much faster when compared to wounding, the authors suggested that the two isogenes are able to sense the difference between touch and wound stimuli.

In addition to the response to mechanical forces, mechanical impedance resulting from pushing against a physical barrier also can accelerate ethylene biosynthesis in plants, which is particularly relevant to growing roots (Hyodo 1991; Morgan and Drew 1997). When roots of maize (*Zea mays*) seedlings were subjected to mechanical impedance, hypoxia, or a combination of both, ethylene production of seedlings as well as activities of ACS and ACO in root tips and ACS in coleoptiles increased (He et al. 1996). A faster and more pronounced response of ACS to mechanical impedance when compared to hypoxia and a strong synergistic effect of both stresses was observed. Considering that the response of ACO was slower but heavier than that of ACS, the authors suggested that enhanced ACS activity probably initiated the acceleration and enhanced ACO activity further contributed to the continuous large increase of ethylene biosynthesis during the course of experiment.

5.2.2 Water Deficit Stress

Increased ethylene production can be found with detached fruits when exposed to (Hyodo 1991) and also after having experienced water deficit during fruit development (Gelly et al. 2003). The transcription of ACS and ACO genes may respond to the water deficit (Fig. 5.1) in a very tissue-specific but coordinated manner as recently described by Nakano et al. (2002). Keeping of persimmon (*Diospyros kaki*) fruits in ambient low humidity conditions (40–60% RH) in comparison with high humidity induced two peaks of ethylene production at days 1–2 and at days 6–8 after harvest. In the calyx, the initial peak of ethylene production was 10-fold higher when compared to the pulp tissue and was associated with accumulation of ACC and induced expression of the ACS gene DK-ACS2, whereas in the pulp no accumulation of ACC and expression of any ethylene biosynthetic genes were detected. In contrast, the second peak of ethylene production was limited to the pulp tissue, associated with increased expression of DK-ACS1, DK-ACS2 and DK-ACO1, and suppressed by application of MCP, which also decreased mRNA abundance of DK-ACS2 and DK-ACO1. The results indicate that ethylene production in response to water deficit was predominantly initiated in the calyx tissue and subsequently induced in the pulp by autocatalysis (Nakano et al. 2002).

Morgan and Drew (1997) already pointed out that the promotion of ethylene production by rapid desiccation of detached leaves cannot be simply applied to intact plants since it is highly influenced by the rapidity of water loss and particular situation of detachment (see Sect. (5.)2.1.1). Reviewing the variable responses described in the literature, the authors concluded that the production of ethylene very much depends on the rapidity of the fall in plant water potential (Ψ), the duration of the stress, and the recovery conditions. This complexity has been further supported and even suggests that the ethylene response might dependent on other related plant or environmental factors. Study on tomato by Kalantari et al. (2000) showed that the data obtained from detached leaves might be absolutely contrary to the response of the whole plant. A rapid dehydration of detached leaves, which decreased leaf Ψ from −0.4 MPa to −1.3 MPa, was inversely related to the continuous increase in the ethylene production rate. In contrast, withholding water from whole plants for a 5-day period resulted in a transient increase in ethylene evolution showing peak between −0.32 MPa to −0.45 MPa. A further decrease in Ψ suppressed ethylene production, below −1.1 MPa even falling below the initial ethylene production of the unstressed plants. Because the decrease in ethylene production was despite the concurrent increase in ACC level and the ethylene production could be dramatically enhanced by re-watering of excised leaves, Kalantari et al. (2000) concluded that ethylene production at low Ψ was limited by a low activity of ACO. This may be caused by a reversible disturbance of functional integrity of plant cells (Fig. 5.1). Thus, extensive cell wall folding and other cellular and metabolic changes in response to strong dehydration of wheat seedlings were associated with decreased ACO activity (Corbineau et al. 2004). In contrast, mild water deficit in sunflower leaves was shown to stimulate expression of ACO genes (Ouvrard et al. 1996). Increasing temperature can promote ethylene production up to a limit of about 35–40 °C, whereas even higher temperatures were shown to inhibit ethylene biosynthesis (Abeles et al. 1992; Verlinden and Woodson 1998; Ketsa et al. 1999; Peirera-Netto et al. 1999; Suzuki et al. 2001; Balota et al. 2004). Considering that strong water deficit impairs temperature regulation of the leaf, high interactions can be expected between air temperature and water deficit.

There are indications that ethylene production can particularly increase after re-watering of previously water-stressed plants and an intermittent accumulation of ACC in roots and subsequent export is probably involved (Fig. 5.1). When well-watered mandarin plants (*Citrus reshni*) were transplanted into dry sand, it took 24 h and a strong decline of leaf Ψ from −0.5 MPa to −3.0 MPa before ethylene production could be detected and afterwards increased in a linear manner (Tudela and Primo-Millo 1992). Whereas ACC concentrations in leaves constantly remained at low levels, a strong increase in ACS activity and ACC level was found in the roots from the beginning of water deficit. When plants were re-watered after 24 h, ACC levels and activity of ACS in roots strongly decreased, whereas a sharp and transient rise in ACC level and ethylene production was observed in the leaves. In a following study on the same cultivar, Gomez-Cadenas et al. (1996) showed that chemical inhibiting of ABA synthesis strongly reduced the water deficit and induced ACC production in roots as well as the subsequent ACC accumulation and ethylene production in leaves after re-watering. Since water-deficit caused an early rise in ABA levels in roots even before ACC accumulation, the authors speculated that ABA induced the ACC production (Fig. 5.1).

When water was withheld for 7 days from wheat plants (*Triticum aestivum*), progressive water deficit only transiently raised the ethylene emission of ears compared to the control plants until the flag leaf Ψ reached −1.0 MPa compared to −0.5 MPa for the unstressed plants (Beltrano et al. 1997).

Further dehydration up to −1.9 MPa gradually decreased the ethylene evolution below the levels of the control. After the plants had been re-watered, ethylene evolution continuously increased above the level of the controls within a 7-h period, which allowed complete recovery of leaf Ψ. A similar kinetics of ethylene evolution in wheat plants in response to strong water deficit and re-watering was also found by Balota et al. (2004) using non-invasive, photo-acoustic trace gas detection. Beltrano et al. (1997) assumed that the concomitant increase of free ROS was causally involved in the pronounced increase in ethylene evolution after re-watering (Fig. 5.1). This hypothesis was supported later by partial suppressing the drought-induced ethylene production of the same wheat cultivar by application of free radical scavenger (Beltrano et al. 1999). However, even though the water-deficit and re-watering schedule was adjusted to same leaf Ψ as in the earlier study (Beltrano et al. 1997), ethylene emission of wheat plants in this study mainly occurred during the water-deficit and was significantly lower after re-hydration. In another study with two other spring wheat cultivars, withholding water induced a transient decrease and subsequent strong increase of ACC and MACC levels in leaves after 24 and 48 h, respectively, which was associated with a continuous rise in ACS activity (Chen et al. 2002). However, only one cultivar responded with a transient increase in ethylene evolution, whereas a decreased ethylene production was generally found after prolonged water deficit and also after subsequent re-watering for 24 h, which partially restored the water content of leaves. Withholding of water to seedlings of winter rye (*Secale cereale*) induced ethylene evolution during the water deficit, which caused wilting symptoms, whereas ethylene production decreased after plants had been minimally re-watered (Yu et al. 2001). Variable responses of the ethylene biosynthetic pathway during the course of changing water availability may be caused by specific transcriptional responses of different genes within same gene families. Such specific responses were recently detected when *Arabidopsis* seedlings were exposed to a short-term water deficit for 3 h (Wang et al. 2005). Activity of GUS-reporter genes revealed a substantial decrease in expression of the ACS gene ACS5 after this treatment, whereas only a slight decrease of ACS7 was detected and ACS4 was not affected at all. There is also indication from *Arabidopsis* mutants (Rao et al. 2002, see Sect. (5.)4.5) that signaling of salicylic acid (SA) is involved in water-deficit induced ethylene production (Fig. 5.1).

5.2.3 Salinity

The influence of salinity on ethylene biosynthesis is even more complicated, since it involves two factors, an osmotic component due to the decreased soil water potential causing secondary water deficit (Sect. (5.)2.2), and a salt-specific component that results from the progressive accumulation of toxic ions in the plant tissues (Munns 1993). This explains the variability of described responses of ethylene biosynthesis to salinity, obviously depending on the particular plant, salt and other experimental conditions.

Salinity may affect the plant during seed germination already. When wheat seeds of two cultivars were germinated in Petri plates, increasing levels of salinity of either Cl- or SO_4-type did not accelerate ethylene evolution by seedlings determined after 6 days even though dry matter of seedlings was strongly reduced (Datta et al. 1998). In contrast, germination of lettuce (*Lactuca sativa*) seeds of nine cultivars in 150 mM NaCl compared to distilled water increased ethylene evolution (Zapata et al. 2003). When seeds of seven plant species were germinated under similar salinity conditions, this induced enhanced ethylene production only with four species including lettuce, whereas even a decrease in ethylene evolution was found with three other species (Zapata et al. 2004). The different responses could not be explained by the ACC levels suggesting that plant specific responses of ACO activity were involved. Exposing rice plants (*Oryza sativa*) of four cultivars to increasing concentrations of NaCl up to 50 mM in nutrient solution raised ethylene production of leaves and ACC concentration in same organs, while the responses were highly variable among cultivars, duration of salt stress and leaf age (Lutts et al. 1996). In tomato plants, ethylene evolution of petioles, and ACC levels in roots, petioles, leaves as well as in premature and mature fruits increased with increasing intensity of NaCl-induced salinity of the nutrient solution (Botella et al. 2000; El-Iklil et al. 2000, 2002).

When rootstocks of *Citrus* were exposed to increasing levels of chloride up to 48 mM using Mg, Ca and K salts, highest chloride level caused a dramatic increase in ethylene evolution of leaves only in the chloride-sensitive genotype, which was associated with an increased ACC level in leaves at 16 mM chloride (Bar et al. 1998). This response could be strongly suppressed by the addition of nitrate, which competes with the chloride uptake. Gomez-Cadenas et al. (1998) found that salt shock to seedlings of a *Citrus* sp. by the application of 200 mM NaCl increased ACC levels in roots, in xylem fluid and in leaves, and dramatically raised leaf ethylene production. The pattern of ACC followed a two-phase response. This was characterized by an initial transient increase, which was obviously related to the osmotic component as reflected by a decrease in leaf and osmotic potential, and an overlapping gradual and continuous accumulation, apparently caused by the accumulation of chloride in leaves. The authors assumed that the osmotic stress-induced and obviously root-derived increase in ACC levels in leaves was caused by the observed increase in root ABA (Fig. 5.1), whereas the second accumulation of ACC was probably originated in leaves. However, in a following study, application of 10 µM ABA to the nutrient solution 10 days before exposure of the rootstock of the same cultivar to 100 mM NaCl significantly reduced the salinity-induced ethylene production (Fig. 5.1) in leaves of the grafted cultivar (Gomez-Cadenas et al. 2002). Because the same treatment also suppressed chloride accumulation in leaves, the authors suggested that ABA might have reduced uptake and translocation of chloride by reducing the stomatal conductance.

Ke and Sun (2004) focussed on the interaction between ROS and ethylene in etiolated mung bean (*Phaseolus radiatus*) seedlings under and following osmotic stress caused by exposure to 25% polyethylene glycol (PEG) for 10 h. When seedlings recovered from the osmotic stress and were compared to the unstressed controls, concentrations of the superoxide radical and H_2O_2 strongly accumulated within 2 h (Fig. 5.1). This was followed by a dramatic increase of ethylene production. Chemical inhibition of ethylene production or perception did not influence the responses of ROS, and application of neither H_2O_2 nor catalase did influence ethylene production. In contrast, applications of generators of superoxide could enhance ethylene production, which was inhibited by exogenous scavengers of superoxide radicals, suggesting that superoxide but not H_2O_2 was involved directly in osmotic-stress-inducible ethylene biosynthesis (Fig. 5.1). However, regarding the decrease of ethylene production when superoxide was provided over extended periods of time or in high concentrations, the authors proposed a dual role of this radical in stress-induced ethylene biosynthesis.

Salinity induced the transcription of ACS isogenes (Fig. 5.1), suggesting that there exist specific responses of the different gene family members, which obviously depend on salt concentration, tissue and other environmental factors. Thus, when leaves of *Arabidopsis thaliana* were treated with different salt solutions, expression of the ACS genes ACS1, ACS3, ACS4 and ACS5 did not respond to either 100 mM NaCl, 50 mM LiCl, or 500 µM $CuCl_2$, whereas ACS6 and ACC level were highly responsive to all salts and ACS2 was only slightly responsive to NaCl (Arteca and Arteca 1999). In contrast, when *Arabidopsis* seedlings were exposed to 300 mM NaCl, GUS-reporter gene activity indicated induced expression of the ACS5 and also of the ACS7 gene (Wang et al. 2005).

5.2.4 Flooding/Hypoxia

Flooding can be defined as any situation of excess of water and may reach extreme situations such as waterlogging, defined as saturation of the root surrounding soil with water, and submergence when the plant is completely covered by water (Peeters et al. 2002). Considering that gases diffuse approximately 10,000 times slower in water than in air (Jackson 1985), excess of water is not only relevant to the natural environment such as river forelands but also to plant production particularly in hydroponic systems. Impaired gas exchange with the atmosphere may cause hypoxia, a condition under which availability of oxygen becomes a limiting factor for ATP production (Dat et al. 2004), and that may occur within plant roots even in well-oxygenated surroundings, when oxygen consumption outpaces the rate of supply of O_2 to the respiring cells (Drew et al. 2000). In addition, the physical enclosure of plant-derived gases in the plant is typical for flooding. Thus, in flooded plant parts, ethylene concentrations may increase due to the physical entrapment even when rate of biosynthesis is unchanged or reduced.

After a positive root-derived signal in waterlogged tomato (Jackson and Campbell 1976) was identified as ACC by Bradford and Yang (1980), following studies completed the picture that roots under anoxic conditions sufficiently export ACC via the xylem to raise the ACO-dependent ethylene production rate in the shoot (Wang and Arteca 1992; English et al. 1995; Else and Jackson 1998). However, restriction on root aeration can stimulate production and accumulation of ethylene in roots if a small amount of oxygen (e.g., 3–5 kPa) remains to support the oxidation of ACC (Jackson and Armstrong 1999). This may be based on increased activity of both ACS (and particularly ACO) in roots as shown for maize seedlings grown at 4% oxygen when compared to normoxic conditions (He et al. 1996, see Sect. (5.)2.1).

During the last decade, evidence has been provided that flooding stimulates the expression of genes for ACS and ACO (Fig. 5.1). Shiu et al. (1998) showed that ethylene production in leaves of waterlogged tomato plants exhibited oscillation and peaks coincided with the light periods. They found that transcription of the ACS gene LE-ACS2 was induced in roots after flooding and then fluctuated peaking during the dark periods. In addition, transcripts of LE-ACS3 strongly accumulated in the same tissue but thereafter constantly remained at high levels. Expression of the gene LE-ACS7 was induced first and preceded the initial peak of leaf ethylene production, whereas none of all these transcripts were present in leaves of flooded plants. The authors suggested that expression of ACS7 has an early and transient function in flooding, possibly also triggering the expression of ACS2 in roots and the ACO in leaves via enhanced ethylene production (Shiu et al. 1998). Soil flooding increased the ACO activity in petioles of a wild-type tomato within 6 to 12 h, which was associated with higher rates of ethylene production, whereas both responses were reduced in a transgenic tomato line coding an anti-sense construct to the ACO gene LE-ACO1 (English et al. 1995). Waterlogging of potato seedlings not only caused a rapid and transient induction of ACO1 and ACO2 in roots but also increased expression of ACO1 in leaves, starting 2 h after flooding and continuing to increase up to 15 h (Nie et al. 2002).

Banga et al. (1996a) found that complete submergence of two *Rumex* species inhibited ethylene release of plants whereas internal ethylene concentrations rapidly increased. Considering that submergence also stimulated accumulation and particularly conjugation of ACC in plant tissues, the authors suggested that submergence stimulated ACC formation and inhibited ACC oxidation to ethylene, which, nevertheless, raised endogenous ethylene levels due to physical entrapment. Very recently, Rieu et al. (2005) confirmed the proposed increased ACC formation for *R. palustris* by analyses of in vitro ACS activity. However, despite the tremendous increase in ACS activity in shoots at 6 h upon submergence, the level of mRNA of the ACS gene RpACS1 in roots was not increased at all and in shoots did not accumulate unless 24 h upon submergence. Even though activity of other ACS genes could not be excluded, the authors suggested that the early increased ACS activity was regulated at post-transcriptional level (Fig. 5.1). By contrast, the transcription of the genes RpACO1 and PpACO2 coding for ACO in *R. palustris* is rapidly and strongly induced upon submergence especially in petioles and is assumed to counterbalance the reduced enzyme activity caused by hypoxia (Vriezen et al. 1999; Peeters et al. 2002).

Submergence not only entraps but also stimulates production of ethylene in deepwater rice by increased activities of ACS and ACO (Métraux and Kende 1983; Cohen and Kende 1986; Mekhedov and Kende 1996). Anoxia induced accumulation of transcripts of the ACS genes OS-ACS3 and OS-ACS1 in roots and shoots, respectively (Zarembinski and Theologis 1993). Submergence induced the expression of OS-ACS1 (Zarembinski and Theologis 1997) and of OS-ACS5, the latter of which may account for the early accumulation of ACC (Zhou et al. 2001).

5.2.5 Chilling

Chilling stress occurs when chilling-sensitive plants are exposed to temperatures ranging from freezing to 12 °C (Saltveit and Morris 1990). Such low temperatures have a great influence on ethylene production of plants. However, the magnitude and even direction of response of ethylene biosynthesis during and following exposure to cold is variable, depending on plant species, cultivar and level of temperature. As already discussed by Morgan and Drew (1997), chilling ethylene may be primary cold-induced or secondary result from other cold-induced primary stresses such as injury (see Sect. (5.)2.1) or water deficit (see Sect. (5.)2.2).

Chilling of fruits may induce a subsequent rise in ethylene production after transfer to higher temperatures (Hyodo 1991; Ben-Amor et al. 1999) but may also enhance ethylene evolution during the low temperature. Thus, ethylene production of pepino fruits (*Solanum muricatum*) raised during storage at 1 °C and coincided with similar responses of free and total ACC levels but was reduced at 10 °C when compared to 20 °C (Martínez-Romero et al. 2003). When fruits of egg plant (*Solanum melongena*) were stored at 10 °C in comparison to 0 °C, only the lower temperature induced high internal ethylene concentrations, which were associated with higher levels of ACC and MACC but contrasted to the concurrently decreasing ACO activity in the pulp (Concellón et al. 2005).

Chilling ethylene may induce the ripening of climacteric fruits of temperate origin (Gerasopoulos and Richardson 1997). Ethylene biosynthesis has been intensively studied in Passe-Crassane pears (*Pyrus communis*), which require a 3-month chilling treatment to be able to ripen autonomously after subsequent re-warming (LeLièvre et al. 1997). Exposure of fruits to 0 °C for 100 days induced high ethylene production and strongly stimulated ACO- and to a lesser extent ACS activity. This was associated with a dramatic accumulation of transcripts of ACO after 40 days and of ACS after 60 days during the chilling treatment (Fig. 5.1). Whereas no such responses were observed when fruits were stored at 18 °C immediately after harvest, fruits which had been previously chilled exhibited a burst of ethylene production upon re-warming, which was parallel to a 1.5-fold stimulation of ACO and even 3-fold stimulation of ACS activity. At the same time, transcripts of ACO (pPC-ACO1) strongly accumulated; however, those of ACS (pPC-ACS1) decreased and reached a minimum after 9 days, which contrasted to the maximum ethylene production. Overnight treatment of fruits with MCP after 27 days of chilling strongly reduced but did not suppress the accumulation of ACO and ACS transcripts. It further caused a rapid disappear of ACO mRNA levels after re-warming, which led to a collapse in ethylene production. A 5-week treatment of non-chilled fruits with the ethylene analogue propylene induced expression of ACO and ripening, whereas ethylene synthesis as well as activity and mRNA levels of ACS remained at very low levels. LeLièvre et al. (1997) concluded from the study that in Passe-Crassane pears ACO gene expression can be induced separately by either chilling or ethylene, whereas expression of ACS requires a cold-related signal and is regulated by ethylene only during or after the chilling treatment.

Marked cultivar differences in the response of ethylene biosynthesis pathway in fruits to chilling were shown for apple (*Malus domestica*) by Larrigaudiere et al. (1997). When fruits of three cultivars were stored at 1 °C, ACC concentrations remained at low levels except for the cv. 'Granny Smith', which strongly accumulated ACC when compared to storage at 20 °C. However, this did not raise ethylene production during the persistent cold, which was limited by the generally low activity of ACO. Re-warming of the chilled fruits to 20 °C raised ACO activity and ethylene production, which in case of 'Granny Smith' strongly surpassed the levels of fruits continuously stored at 20 °C, while the same cultivar showed an outstanding decrease in ACC levels. Focusing on this responsive cultivar, a detailed study of Lara and Vendrell (2003) revealed remarkable differences in the cold-induced ethylene biosynthesis in peel and pulp tissue of the fruit. The kinetics of ethylene production, ACO activity, ACC levels, and accumulation of ACO and ripening-related ACS proteins during and upon removal from cold-storage, and the responses to exogenous ethylene indicated that increased ACO activity in chilled fruit after re-warming resulted from both enhanced biosynthesis (peel) and activation (pulp) of the protein and caused increased ethylene levels, which enhanced ACS activity leading to a rapid onset of ethylene biosynthesis. Because application of ABA restored ACO activity and enhanced ACC concentrations in the peel during cold storage, and chilling significantly raised endogenous ABA levels in the same tissue upon re-warming (Fig. 5.1), Lara and Vendrell (2003) suggested that increased endogenous ABA levels in the peel tissue might play a major role in cold-induced ethylene biosynthesis upon re-warming, possibly via activation of the ACO protein.

Different genes of the same ACS multi-gene family may differently respond to chilling. This was demonstrated by Wong et al. (1999) with non-climacteric citrus fruits (*Citrus sinensis*), which showed de-greening and yellowing of the peel in response to both low temperatures and application of ethylene. Storage of fruits at 4 °C when compared to 25 °C induced a dramatic accumulation of transcripts of the CS-ACS1 gene in the peel but reduced the constitutive expression of the CS-ACS2 gene below detection level. However, both genes, which were also wound-inducible, showed a rapid induction of transcripts upon subsequent re-warming. After 24 h of re-warming, CS-ACS-1 diminished to undetectable level, whereas the CS-ACS-2 mRNA regained its basal level of expression attained prior to the chilling treatment. Ethylene production was diminished after 3 and 6 days of cold storage. ACC level was lower after 3 days and was significantly higher after 6 days of chilling and also was correlated with the increased expression of CS-ACS1. After re-warming, both ACC levels and ethylene production increased compared to the non-chilled control, while higher peak levels were obtained with the longer chilling period. Wong et al. (1999) concluded that the expression of both the chilling-inducible and the chilling-repressible ACS genes play an important role in determining the level of ethylene production in citrus peel during the post-chilling period, whereas CA-ACS1 is highly responsible for elevating the ACC level during chilling. In a following study, it was shown by the same group that the chilling-induced ACC accumulation in branch tissues of *Citrus sinensis*, *Carrizo citrange*, and *Poncirus trifoliate* can be decreased by over-expression of antisense ACS-1 mRNA (Wong et al. 2001).

Chilling also influences ethylene biosynthesis of vegetative tissues. Monitoring the ethylene emission of pelargonium cuttings (*Pelargonium x hortorum*) with on-line photoacoustic detection revealed, that cold storage at 3 °C of excised cuttings suppressed production of wound ethylene, but induced subsequent production of chilling ethylene upon re-warming (Kadner et al. 2000). The influence of short-term chilling on the expression of individual members of the ACS and ACO families was recently analysed in vegetative tissues. When seedlings of transgenic *Arabidopsis* lines expressing the GUS and GFP reporter genes from the promoter of nine different ACS gene family members were incubated at 4 °C for 24 h, this inhibited the expression of ACS5 and ACS11 and altered the pattern of ACS8 expression in roots when compared to 25 °C (Tsuchisaka and Theologis 2004). Nie et al. (2002) studied the expression of the ACO genes ACO1 and ACO2 in potato (*Solanum tuberosum*). Incubating tubers with small sprouts at 0 °C for 24 h stimulated the expression of both ACO genes in sprout and tuber tissues when compared to moderate temperatures, while the same genes were also wound-inducible in leaves.

5.2.6 Ozone

Exposure of plants to air pollutants can enhance ethylene production (Hyodo 1991; Morgan and Drew 1997). Since ozone (O_3) is a major phytotoxic component, intensive studies have been focussed on O_3-induced ethylene biosynthesis and its role in plant damage (see Sect. (5.)4.2). Exposure of potato plants to O_3 induced ethylene and ACC production in leaves, which was associated with a sequential expression of the two ACS genes, ACS4 (slow) and ACS5 (rapid) (Schlagnhaufer et al. 1997). Similar responses were also found with the application of copper, another inducer of oxidative stress, and after infection with *Alternaria solanii*. The sequential nature of expression of both ACS genes indicated two different signal transduction and gene regulatory mechanisms, which might be interconnected through an autocatalytic process induced by the early responding ACS5 (Schlagnhaufer et al. 1997). A similar sequential expression of rapid and slow ACS was also found in tomato by Nakajima et al. (2001). When seedlings were exposed to O_3, ethylene became detectable after 1 h and increased almost linearly until 4 h. This was paralleled by raised ACS activity and ACC concentrations detected after 1 h, reaching maximum levels after 2 h, and declining after 4 h. The early rise of ACS activity was associated with accumulation of transcripts for LE-ACS1a and LE-ACS6, which, however, after peaking at 2 h, declined to pre-stress levels until 6 h. An increased expression of LE-ACS2 was detected earliest after 2 h but then was kept at the high level thereafter. O_3 also increased the level of the ACO transcript ACO1 as early as 1 h after the start of exposure. The level increased linearly up to 4 h and still was high after 6 h. Expression of all three ACS isozymes was also induced by wounding of leaves (Tatsuki and Mori 1999, see Sect. (5.)2.1.2). However, the transcript of a wound-inducible proteinase inhibitor II was undetectable during ozone treatment, which did not support an involvement of wound stress in O_3-induced ethylene biosynthesis (Nakajima et al. 2001). Regarding the kinetics of all parameters, the authors suggested that ACSs initially regulated ethylene biosynthesis at the early stage of ozone exposure and LE-ACO1 limited ethylene production after ACC was sufficiently accumulated.

5.3 Stress-Mediated Ethylene Sensitivity

There is increasing evidence that stress not only acts via stimulation of ethylene biosynthesis but also can modify ethylene sensitivity of plant tissues. This is first supported by changed phenotypic responses to certain doses of ethylene indicating that exposure to stress may increase but also decrease tissue sensitivity to ethylene. The senescence response of petals from *Portulaca* hybrids to ethylene was significantly increased within 1 h after wounding of filaments (Ichimura and Suto 1998). There is also indication that leaf senescence in rice in response to water-deficit and osmotic stress is not primary mediated by increased ethylene production but rather by increased sensitivity to this hormone and NH_4 accumulation may be causally involved (Lin and Kao 1998; Hsu et al. 2003). The role of ethylene biosynthesis and perception in chilling-induced leaf abscission of *Ixora coccinea* plants was intensively studied by considering the interaction with oxidative stress and auxin economy (Michaeli et al. 1999a, 1999b). Exposure of plants to chilling temperatures (3–9 °C for 3 days) caused increased leaf abscission after transfer to 20 °C. This was preceded by an increase in ethylene production rates of the abscission zone (AZ) tissue during the first 5 h of the re-warming, which obviously resulted from the ACC accumulation during the chilling period. However, treating plants prior to chilling with antioxidants significantly reduced chilling-induced leaf abscission despite the chilling-induced ethylene production remained unaffected. Furthermore, exposure of plants to ethylene enhanced leaf abscission only when they had been chilled, whereas treatment with MCP inhibited both the chilling-induced and ethylene-enhanced leaf abscission. Chilling reduced free IAA concentrations in the AZ and leaf blade, enhanced the decarboxylation of IAA, particularly in the AZ zone, and inhibited auxin transport capacity in the petioles. The chilling-induced leaf abscission was almost completely inhibited by the synthetic auxin α-naphthalenacetic acid (NAA), while application of antioxidants reduced the decline in free auxin levels and auxin transport capacity. These results strongly suggest that i) ethylene is essential for chilling-induced abscission in *Ixora* but not the triggering factor, and ii) chilling-derived oxidative processes trigger leaf abscission via increased sensitivity to ethylene, which was probably caused by a reduced availability of free IAA in the AZ zone (Michaeli et al. 1999a, 1999b). Contrarily, treatments with the ethylene-releasing compound, ethephon and ethylene action inhibitor, STS indicated that cold storage of cut lilies (*Lilia* hybr.) decreased the sensitivity to ethylene in terms of flower abscission or bud abortion (Song and Peng 2004).

Contrasting to the well-known biosynthetic pathway, perception of ethylene freely diffusing through the membranes and cytoplasm and further downstream signal transduction is still far from complete understanding. However, after major elements have been first identified and functionally analysed in the model plant *Arabidopsis thaliana*, subsequent investigations on other plant species suggest a signaling cascade which is probably highly conservative in the plant kingdom (Alonso and Stepanova 2004; Klee 2004). The actual mechanistic model involves a family of endoplasmatic reticulum (ER)-localized ethylene receptors that share sequence similarity with the bacterial two-component histidine kinases. The high binding activity and specificity to ethylene is achieved with the help of a copper cofactor associated with the hydrophobic domain. In *Arabidopsis*, five ethylene receptors, ETR1, ETR2, EIN4, ERS1, and ERS2 have been characterized. According to the model, the receptors are active in the absence of ethylene and suppress the ethylene response by stimulating the negative regulator CTR1, a Raf-like serine-threonine kinase, which in turn shuts off the ethylene-signaling pathway (Fig. 5.2). Binding of ethylene to the receptors inactivates them and thereby relieves this CTR1-mediated blockage of further positive downstream signaling events. The further downstream cascade probably includes the positive

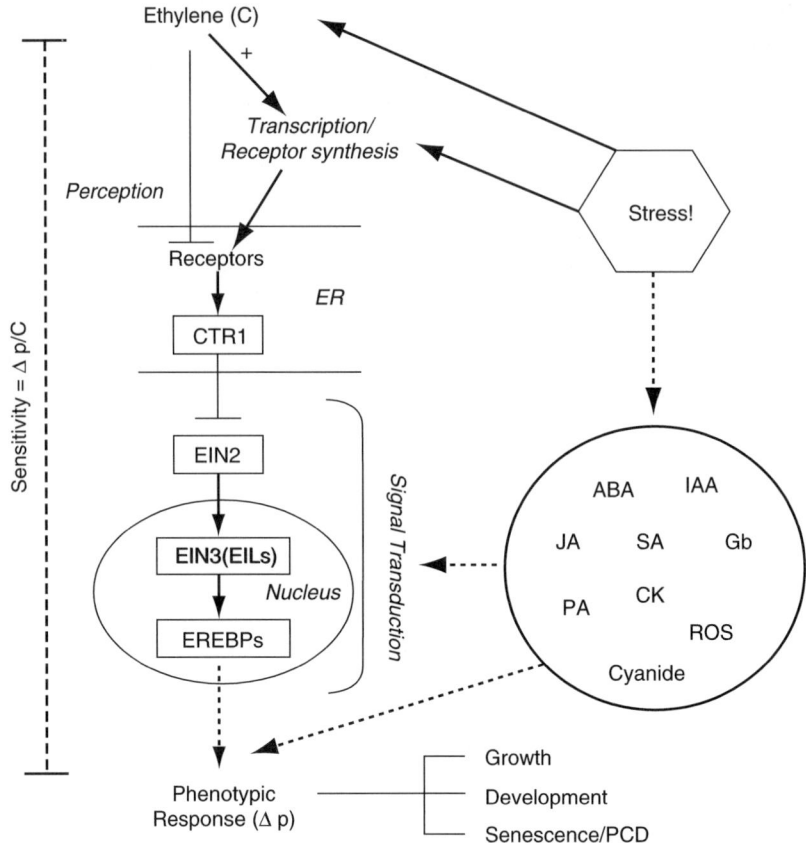

Fig. 5.2. A schematic model illustrating the relationships between stress, ethylene and the phenotypic response. The model integrates the *Arabidopsis*-based ethylene signalling pathway (Alonso and Stepanova 2004; Guo and Ecker 2004), includes the plant hormones JA, IAA, ABA, SA, and PA and the compounds ROS and cyanide discussed in Sects. 5.3–5.4, and also considers the plant hormones cytokinins (Ck) and gibberellins (Gb) according to Gazzarani and McCourt (2003). Stress influences the ethylene concentration (C) and the formation of ethylene receptors, directly and/or via ethylene, the relation of which determines ethylene perception and subsequent signal transduction. Stress also influences the concentration and/or action of other plant hormones and compounds, which may be partially mediated by ethylene (Fig. 5.1). These compounds may interact with elements of the primary ethylene signalling pathway, like JA with ERF1, or may be integrated later to influence the phenotypic response (Δp), which determines the stress-mediated ethylene sensitivity ($\Delta p/C$)

regulatory molecule EIN2, transmitting a positive signal to the transcription factors, EIN3 and EILs (EIN3-like genes) in the nucleus, resulting in accumulation of EIN3/EIL proteins, which then induce the transcription of genes for ethylene responsive element binding proteins (EREBPs), such as the ethylene response factor ERF1 (Guo and Ecker 2004).

Interestingly, ethylene itself was shown to induce the expression of the receptors ETR2, ERS1, and ERS2 (Fig. 5.2), but not of ETR1 and EIN4. According to this model, an increased synthesis of ethylene receptors can be expected to de-sensitize the plant tissues for ethylene because more ethylene molecules are necessary for inactivating the receptor-controlled blockage of the signaling cascade. Recent studies indicate that exposure of plants to stress may change transcription of genes evidently or putatively involved in ethylene perception (Fig. 5.2). However, the information on this topic is still fragmentary and confusing. Voesenek et al. (1997) demonstrated by using inhibitors of ethylene synthesis and action that hypoxia enhanced petiole extension in *R. palustris* by increasing the sensitivity to ethylene. However, the expression level of a putative ethylene receptor in *R. palustris,* Rp-ERS1 was up-regulated by hypoxia and flooding, which did not accord with the observed increased sensitivity and the model of a constitutively active receptor (Voesenek et al. 1997; Vriezen et al. 1997). Complete submergence of seedlings of deepwater rice did not alter expression of the gene OS-ERS1 coding for a putative ethylene receptor (Zhou et al. 2001) but induced the transcription of OS-ERL1, another ethylene receptor homolog (Watanabe et al. 2004). Gene expression profiling in *Arabidopsis thaliana* seedlings using whole-genome DNA amplicon microarrays (Liu et al. 2005) revealed that transcript levels of genes coding the ethylene receptor ETR2 and a specific EREBP were increased in response to hypoxia.

Expression of other receptor homologues responded to wounding, salt stress, chilling and exposure to O_3. Wounding of peach fruits caused an increase of transcripts of the gene PpETR1, a peach homologue of ETR1 whereas such response was not detected in leaves (Bassett et al. 2002). In wheat seedlings, wounding as well as application of either JA or ABA increased the expression of an ethylene receptor homologue, W-ert1, which predicted amino acid sequence is over 70% similar to ERS1 from *Arabidopsis* (Ma and Wang 2003). Expression of a gene for a putative ethylene receptor in rice, OSPK2, which was found to be divergent from homologues in dicots and assumed not to function as a histidine kinase, could be induced in the shoot by wounding and in shoots and roots by treatment with PEG (Cao et al. 2003). Zhang et al. (2001) demonstrated that the two-component gene NTHK1 (*Nicotiana tabacum histidine kinase*-1), which shared a high homology with the LETR4 from tomato and encoded a putative ethylene-receptor homolog, was inducible in tobacco upon wounding as well as following treatment with NaCl or PEG. The constitutively suppressive action of NTHK1 on ethylene signaling became evident by reduced ethylene sensitivity in transgenic *Arabidopsis* over-expressing of this gene (Xie et al. 2002). In a following study, He et al. (2004) analysed the expression of a gene encoding for a NTHK1-regulated receptor-like kinase, AtLecRK2, in response to 200 mM NaCl. They found that salt stress induced the transcription level of AtLecRK2 in wild-type *Arabidopsis* whereas induction was inhibited and retarded in the transgenic NTHK1 plants. AtLecRK2 was constitutively over-expressed in the ethylene-overproducer mutant, *eto1-1*, and could be induced by the application of ACC. The results indicate that the induction of ATLecRK2 in response to salt stress is regulated by the ethylene signaling via deactivation of the ethylene receptor, NTHK1, constitutively suppressing the expression of ATLecRK2 (He et al. 2004). The missing salt response of ATLecRK2 expression in the ethylene-insensitive mutant *ein2-1* indicates that expression of ATLecRK2 may need only the components of the signaling pathway upstream of EIN2.

El-Sharkawy et al. (2003) isolated four cDNAs from pear (*Pyrus communis*), which were homologous to the ethylene receptor genes ETR1, ERS1, and ETR2 and to the negative regulator gene CTR1 in *Arabidopsis,* and were designated Pc-ETR1a, Pc-ERS1a, Pc-ETR5, and Pc-CTR1, respectively. Chilling of Passe Crassane pears induced accumulation particularly of Pc-ETR1 and also of Pc-ETR5 transcripts during the cold period of 0 °C. When fruits were removed from the cold and stored at 20 °C, this resulted in an increase of transcripts of all genes studied, which was not observed without previous chilling. The peak of Pc-ETR5 and Pc-CTR1 coincided with the climacteric peak of ethylene production. Both increases of gene expression during and following the cold treatment were dependent on ethylene action, since inhibition of ethylene perception with MCP resulted in elimination of any increase (El-Sharkawy et al. 2003). The authors suggested that the increased receptor expression during the cold treatment contributed to the attainment of ripening capacity by counteracting the minor increase in ethylene production, and the pronounced increase of receptors after re-warming altered the ripening response to the concurrent burst of ethylene production (see Sect. (5.)2.5.1).

Zhao and Schaller (2004) recently analysed the expression of the evident ethylene receptor ETR1 in *Arabidopsis* in response to salt and osmotic stress. In contrast to the papers mentioned above, they found that expression of ETR1 was reduced after exposure to salt (200 mM NaCl) and osmotic stress, which was also reflected at the protein level. The authors concluded that the reduced ETR1 levels should make the plant more sensitive to ethylene provided other receptors respond in a similar way. They further suggested that reduction in receptor levels might contribute to the maintenance of an extended stress response. That the same stress may affect the transcription of individual receptor genes in opposite directions was shown by Moeder et al. (2002). A 5-h pulse of ozone induced accumulation of transcripts for the ethylene receptor LE-ETR1, whereas the transcripts of another receptor, LE-ETR2, decreased markedly.

5.4 Ethylene Action, Adaptation and Stress Tolerance

Levitt (1980) differentiated two adaptation strategies of plants to gain a higher level of resistance, (i) avoiding thermodynamic equilibrium with the stress factor, and (ii) preventing, decreasing or repairing the injurious strain. The term stress tolerance will be used here for both strategies as frequently done in the literature. However, the interpretation of a certain plant response to stress may depend on the perspective. Thus, from an eco-physiological point of view, decreased shoot growth, epinasty and even leaf abscission in response to salt stress, water deficit or flooding may contribute to increased tolerance of the plant because subsequent water loss will be avoided, whereas from an agricultural point of view same responses may be regarded as damage. Nevertheless, ethylene action is highly involved in these processes.

5.4.1 Wound Response and Thigmomorphogenesis

Together with JA, ethylene action is involved in the wound response, which is orchestrated by complex signaling pathways providing tissue repair as well as increased resistance to wound-causing agents such as plant invaders (O'Donnell et al. 1996; Howe 2004). Mechanically stimulated ethylene biosynthesis may contribute to physiological and developmental changes, such as reduced elongation or increased radial expansion of shoots or roots, which enhance resistance to subsequent mechanical forces and are termed as thigmomorphogenesis (Hyodo 1991; Jaffe and Forbes 1993; Morgan and Drew 1997; Sunohara et al. 2002). However, the role of ethylene action in thigmomorphogenesis of *Arabidopsis thaliana* was questioned by Johnson et al. (1998), since the growth responses and up-regulation of TCH (touch) gene expression in response to different mechanical stimulations was also found with the two ethylene-insensitive mutants, *etr1-3* and *ein2-1*.

5.4.2 Ethylene and Ozone-Induced Cell Death

Concentration of ozone (O_3) has increased markedly in the air and causes extensive damage to natural and cultivated plants (Pell et al. 1997). When plants are exposed to O_3 at high concentrations, necrotic lesions appear on leaves reflecting the death of mesophyll cell clusters (Treshow and Anderson 1989). This may be caused by the oxidation of biomaterials via generated ROS (Mudd 1996). However, there is increasing evidence that similar to the pathogen-induced hypersensitive response, O_3-induced activation of programmed cell death (PCD) also underlies such damages and ethylene is involved in the signaling events (Kangasjärvi et al. 1994; Kanna et al. 2003). Tingey et al. (1976) already showed that O_3 induced ethylene production in more than 25 plant species and ethylene emission was positively correlated with the expression of visible injury. The analyses of these relationships particularly in *Arabidopsis* and members of the *Solanaceae* during the last decade provided evidence that accelerated ethylene production was not simply the result but rather a trigger of the O_3-induced damages.

Vahala et al. (1998) investigated the O_3-response of different members of the ACS gene family in leaves of *Arabidopsis thaliana*. Maximum ethylene evolution in response to exposure to high O_3 concentration was followed by visible damage but was preceded by induced transcription of a mRNA, which sequence was found to be identical to that of ACS6, a wound and touch-inducible ACS gene (see Sects. (5.)2.1.2 and (5.)2.1.3). Following studies compared the ethylene over-producer mutants *eto1* and *eto2* with the O_3-tolerant wild-type accession *Col-0* and analysed the biosynthesis of ethylene, SA, and JA as well as the transcription of respectively dependent genes and lesion development in response to O_3, inhibitors of ethylene biosynthesis and exogenous SA. The results strongly support the conclusion that O_3-induced cell death in *Arabidopsis* is the result of a crosstalk between ethylene, SA and JA (Overmyer et al. 2000; Rao et al. 2000, 2002). That SA signaling is required for stress-induced ethylene production in *Arabidopsis* was supported by the finding that plants of the low-SA-transformant *NahG* expressing the gene for SA-degrading salicylate hydroxylase and of the mutant *npr1* blocked in SA-dependent systemic acquired resistance failed to produce ethylene in response to O_3, while water-deficit induced ethylene production was also inhibited in *NahG* plants (see Sect. (5.)2.2). Furthermore, *NahG* expression in the dominant *eto3* mutant attenuated expression of an ethylene dependent PR gene and rescued the O_3-induced cell death exhibited by *eto3* plants. Based on these results, Rao et al. (2002) proposed a model of SA-ethylene- and JA-cross-talk in O_3-induced cell death, which is illustrated in Fig. 5.3. According to this model, O_3, upon entering the leaf tissue via stomata, generates ROS, resulting in increased biosynthesis of signaling molecules such as SA, which in turn potentiates the feedback amplification loop of runaway cell death cycle that induces the biosynthesis of signaling molecules such as ethylene. On the other hand, O_3, either by reacting with membrane lipids or by generating ROS, induces the biosynthesis of JA or MJ, which in an antagonistic manner attenuate development of cell death. However, the authors considered the possibility that SA was required for additional stress-induced components to maximize ethylene production and pointed out that this type of interaction might not hold true for other plant species or other type of stresses.

Tamaoki et al. (2003) compared the physiological and molecular responses to O_3 of the tolerant accession *Col-O* and the sensitive accession *Ws-2*.

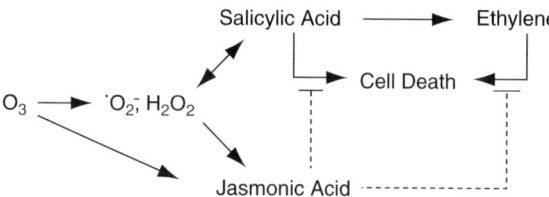

Fig. 5.3. A schematic model illustrating the synergistic and antagonistic interactions between SA-, ethylene-, and JA-signaling pathways on O_3-induced cell death (Rao et al. 2002, with permission)

Stronger lesion development and ion leakage in the *Ws-2* plants was preceded by a more pronounced increase in ethylene evolution and stronger accumulation of ACS6 transcripts, whereas the damage was attenuated by pretreatment with inhibitors of ethylene biosynthesis and perception. RNA blot analyses showed that O_3-induced increases in mRNA levels of several ethylene-inducible genes and the SA-inducible gene PR1 were substantially higher in *Ws-2* compared to *Col-O* plants. Because induction of ethylene-inducible genes was observed prior to SA-related gene induction, Tamaoki et al. (2003) speculated that ethylene and SA might depend on each other to induce O_3-triggered cell death. Analyses of O_3-induced ethylene evolution and leaf injury from 20 *Arabidopsis* accessions revealed four clusters, two of which did not show a relationship between ethylene evolution and leaf damage. This was explained by variable ethylene sensitivity or interaction with JA signalling (Tamaoki et al. 2003). The defensive role of JA signalling by suppression of O_3-induced cell death (Fig. 5.3), possibly via reduction of ethylene biosynthesis, was supported by characterization of a JA-semi-sensitive mutant, *Ojil* (Kanna et al. 2003). The decreased JA-sensitivity of *Ojil* in comparison to wild-type plants, *Ws-2*, was demonstrated by a weaker response of root growth to MJ application and less pronounced O_3-induced expression of the JA-inducible AtVSP1 gene. This mutant responded to O_3-exposure with more severe foliar injury, higher ethylene emission, a stronger increase in expression of the SA-inducible gene PR1, and a stronger accumulation of endogenous JA. With both genotypes, the O_3-induced injury and ethylene emission could be reduced by chemical inhibiting of ethylene biosynthesis. In contrast, pretreatment with MJ at 10 µM suppressed O_3-induced ethylene emission and foliar injury only in *Ws-2* plants, whereas *Ojil* plants responded in the same manner to MJ only when applied at 100 µM.

The limiting role of ACS in O_3-stimulated ethylene biosynthetic rate and leaf damage was supported in a study using transgenic tobacco plants constitutively expressing an anti-sense DNA for the early inducible tomato gene LE-ACS6 (Nakjima et al. 2002). Obviously depending on the copy numbers and/or position of the anti-sense DNA in the host genome, only three of seven transgenic lines showed lower rates of O_3-induced ethylene production when compared to a wild-type. Nevertheless, the extent of leaf injury was generally positively correlated to the level of ethylene evolution. In the most ozone-resistant line, O_3-induced accumulation of ACC and transcription of two sequentially expressed ACS genes, NT-ACS-2 and NT-ACS-6, with nucleotide sequences similar to LE-ACS6 were suppressed compared with those in wild-type plants. The relationship between O_3-induced expression of genes encoding of components of ethylene biosynthesis and perception and cell death in tomato was further highlighted by Moeder et al. (2002), who completed the picture of a biphasic gene regulation and uncovered the cellular distribution of molecular events. Rapid transcription responses to a pulse of O_3 were found after 1 h and included the induction of LE-ACS6, LE-ACO1 and LE-ACO3, which were followed by induction of LE-ACS2, LE-ACO2 and

LE-ACO4 after 2 h. Transgenic tomato harbouring the LE-ACS2 gene in antisense orientation showed lower ACC accumulation and ethylene production only during the second phase after 2 h of O_3 exposure and were not more tolerant to O_3 than the wild-type. Also, reduction of O_3-induced ethylene production by about 50% in transgenic anti-sense LE-ACO1 plants failed to reduce leaf damage. These results suggest the importance of early ethylene biosynthesis and a certain threshold of ethylene production, which has to be passed before stimulating lesion formation (Tuomainen et al. 1997; Moeder et al. 2002). However, strong chemical blocking of ethylene biosynthesis or perception in wild-type plants significantly reduced O_3-induced tissue damage and this was associated with significantly lower H_2O_2 accumulation in leaves. Furthermore, with the aid of transgenic plants expressing the LE-ACO1 promoter:GUS fusion, the authors showed that GUS expression increased rapidly (after 30 min) at the beginning of the O_3 exposure and had a spatial distribution resembling the pattern of subsequent extracellular H_2O_2 production at 7 h, which coincided with the cell death pattern after 24 h. These results strongly suggest that ethylene is intimately involved in the amplification of ROS production by the plant cells (second oxidative burst, Schraudner et al. 1998) and regulation of cell death under oxidative stress (Moeder et al. 2002).

Ethylene synthesis may also contribute to cell death via production of cyanide (Fig. 5.1). Vahala et al. (2003) showed that disruption or blocking of ethylene perception in birch (*Betula pendula*) by transformation with the *etr1–1* gene of *Arabidopsis* or application of MCP, respectively, reduced (but not completely prevented) O_3-induced cell death, whereas chemical inhibition of ethylene biosynthesis completely prevented lesion formation. Functional ethylene signaling was required for O_3-induction of the gene encoding for CAS (Fig. 5.1), suggesting that ethylene signaling contributed to detoxification of cyanide (Vahala et al. 2003).

5.4.3 Ethylene and Root Stress-Mediated Shoot Growth

Considerable attention has been given to the involvement of root-sourced ABA not only in the reduction of stomatal conductance but also of leaf/shoot growth in response to soil compaction and water deficit stress (Davies and Zhang 1991; Tardieu et al. 1992). However, there is increasing evidence that ethylene action is highly involved in these responses. Hussain et al. (1999) investigated the interrelationships between ABA and ethylene in regulation of shoot growth of tomato in compacted soil using the wild-type cultivar Ailsa Craig in comparison with the transgenic tomato line $ACO1_{AS}$, which has a reduced capacity to synthesize ethylene due to anti-sense ACO. When plants were grown in uniformly compacted soil or in a split-pot treatment (non-compacted + compacted soil) and compared with a non-compacted soil treatment, xylem sap ABA levels were similarly increased in both the genotypes, whereas enhanced ethylene production was only found with Ailsa Craig plants. Growing the plants in the split-pot treatment invoked marked genotypic differences in the growth response. Reduction in growth and leaf expansion compared to the non-compacted soil were restricted to the cv. Ailsa Craig. Excising the roots in the compacted compartment reduced ethylene evolution in this genotype, while the same treatment as well as chemical blocking of ethylene action restored shoot and leaf growth. In contrast, treatment with ethephon reduced growth in $ACO1_{AS}$. Also for the ABA-deficient line *notabilis*, reduced growth was observed in the split-pot treatment which was associated with increased ethylene production (Hussain et al. 2000). However, application of ABA had little effect on $ACO1_{AS}$ but promoted a recovery of leaf expansion in *notabilis* and Ailsa Craig. The authors suggested that leaf expansion is probably regulated by an antagonistic interaction between ABA and ethylene, when the root system simultaneously encounters non-compacted and compacted soil.

New studies including ABA-deficient mutants, inhibitors of ABA synthesis and changed experimental strategies also provided a changed view on the role of ABA and ethylene in growth responses to water deficit. Altogether, there are indications that accelerated ethylene biosynthesis is an important early drought-induced signal, which can be suppressed by ABA in an antagonistic manner and obviously depending on the duration of water deficit can improve or alternatively reduce shoot-growth stress (Sharp 2002; Sharp and LeNoble 2002). The importance of ethylene production was also recently highlighted in a split-root system by studying the influence of partial root-zone drying (PRD) on the tomato cultivar Ailsa Craig in comparison with the low-ethylene producing transgenic line $ACO1_{AS}$ (Sobeih et al. 2004). In response to PRD, which did not change leaf Ψ when compared to the non-stressed controls, both genotypes showed similar increases in xylem sap pH, ABA concentration and decreased stomatal conductance, whereas only Ailsa Craig responded with enhanced ethylene evolution of leaves and significant reductions in leaf elongation. Sobeih et al. (2004) suggested that such responses might occur when mild soil drying, in which leaf Ψ is maintained or declines very slowly, stimulates ethylene synthesis (Sect. (5.)2.2). However, it has to be questioned whether a strong dehydration of a part of the root system is a mild stress to the plant. Even though a substrate of low bulk density was used in that study, mechanical stress and root injury (Sect. (5.)2.1) might have been involved.

5.4.4 Ethylene and Adaptation to Salt Stress

Ethylene action may trigger salinity-induced phenotypic responses of plants. Leaf abscission in salt-stressed *Citrus* plants was associated with increased ethylene production and could be reduced by foliar applications of the ethylene action inhibitors $CoCl_2$ or STS (Gomez-Cadenas et al. 1998). However, the role of ethylene in leaf senescence is complex (Sect. (5.)4.2). The competition between ethylene biosynthesis and synthesis of PA for the common precursor, SAM (Sect. 5.2), the possible up-regulation of the key enzyme in PA biosynthesis, SAMDC by ethylene (Hu et al. 2005), and the role of PAs as radical scavengers and stress protectants (Bors et al. 1989) have also to be considered. These interactions became apparent by the finding that improved tolerance of transgenic tobacco plants expressing anti-sense ACS or anti-sense ACO genes to different abiotic stresses including salinity in terms of chlorophyll loss and phenotypic changes was correlated not only with decreased ethylene production but also with raised PA contents particularly with oxidative stress (Wi and Park 2002). According to these results, complicated three-way interactions were found in Glycorrhiza (*Glycorrhiza inflata*) and wheat plants under osmotic stress between the synthesis of ROS, of ethylene and of PA, which were dependent on the degree of stress-induced leaf damage (Li and Wang 2004; Li et al. 2004). Application of ACC promoted stress-induced leaf damage, whereas both blocking of ethylene biosynthesis and application of the PA spermine had opposite effects.

When El-Iklil et al. (2000) compared four varieties of tomato under salinity conditions higher ethylene production rates of petioles were associated with less pronounced epinasty and higher dry matter accumulation. However, the two varieties exhibiting less epinasty and growth reduction revealed a higher basal level but a less pronounced relative increase in ethylene production in response to salinity. These results indicate that the responsiveness of ethylene biosynthesis to stress is relevant to stress tolerance. Such relationships are further supported by the observation that improved drought tolerance of rose plants (*Rosa* x *hybrida*) in terms of reduced water loss and retarded wilting in response to previous acclimation to low water availability was associated with an alleviated rise of ethylene production in leaves during the stress, which probably was caused by the increased conjugation of ACC in roots (Andersen et al. 2004).

There is indication that ethylene action may also increase tolerance of plants to salt stress. When seeds from nine lettuce cultivars were germinated under salt stress, a higher salt-induced increase in ethylene evolution of seedlings was associated with a lower reduction in fresh weight (Zapata et al. 2003). Such a relationship was not evident, when seven plant species were compared to each other under similar conditions (Zapata et al. 2004), which did not exclude a role of ethylene but rather revealed that salt tolerance could not be explained by only one factor. Also when rice plants of four cultivars were exposed to NaCl, those cultivars revealing a higher salt tolerance by shoot growth and tissue viability showed a stronger increase in ACC level and ethylene production rate (Lutts et al. 1996). Furthermore, ethylene production rates under saturation with ACC indicated that a salt-induced reduction of ACO activity was less pronounced in the tolerant compared to the more sensitive genotypes. The role of ethylene action in enhancing salt tolerance was recently supported at molecular level. Wang et al. (2004) reported a novel

ERF protein, JERF3, which unites ethylene, JA and osmotic signaling pathways. Expression of JERF3 in tomato was mainly induced by cold, salt, ethylene, JA or ABA. Constitutive expression of this gene in transgenic tobacco activated expression of pathogenesis-related (PR) genes and enhanced salt tolerance as determined by decreased chlorophyll loss of leaf discs in NaCl solution.

5.4.5 Ethylene and Flooding Tolerance

Epinastic growth of petioles and downward rolling of leaf laminae is a characteristic adaptive response of *Solanaceae* to hypoxia of roots caused by soil flooding, which minimizes evaporative demand to counteract low hydraulic conductivity of anaerobic or even dead roots and is inducible by ethylene (Wang and Arteca 1992; English et al. 1995; Else and Jackson 1998). That ethylene is causally involved in this response became evident from alleviation of ethylene production of petioles and final epinastic curvature of leaves in transformed tomato expressing the bacterial gene ACC deaminase, which probably reduced the ACC pool (Grichko and Glick 2001). Accumulation of ethylene probably accounts for flooding-induced rapid stem elongation (Kende et al. 1998) and aerenchyma formation in roots of plant species well-adapted to waterlogging, which provides new internal long-distance gas transport pathways (Jackson and Armstrong 1999; Drew et al. 2000), and also for induction of adventitious roots, which in case of *Rumex palustris* can be explained by increased sensitivity of the root-forming tissue to endogenous IAA (Visser et al. 1996). Comparison of this flooding-tolerant species with *R. acetosella*, a flooding-sensitive species revealed that only *R. palustris* responded to submergence with stimulated elongation of rosettes, which helps the plant to escape from submergence. This phenotypic response could be explained by different ethylene sensitivity when compared to *R. acetosella* (Banga et al. 1996b; Voesenek et al. 1997).

5.4.6 Ethylene and Chilling Response

Increased ethylene production may cause chilling injury developed by fruits of tropical and subtropical origin during storage or upon re-warming (Hyodo 1991; Morgan and Drew 1997; Martínez-Romero et al. 2003; Concellón et al. 2005). In contrast to wild-type fruits of cantaloupe melon (*Cucumis melo*), fruits of an anti-sense ACO genotype exhibited reduced membrane deterioration and less visual damage of peel tissue during low temperature storage and upon re-warming, which was correlated with very low ethylene production (Ben-Amor et al. 1999). Treatment with the ethylene perception inhibitor MCP prior to cold storage suppressed the chilling symptoms in wild-type fruits, whereas application of 10 ppm ethylene to the ACO anti-sense fruits prior to storage restored the sensitivity to chilling.

Ethylene biosynthesis may also be involved in chilling tolerance of whole plants. Ciardi et al. (1997) compared the Never-ripe (*Nr*) tomato exhibiting a decreased ethylene sensitivity with a normal isogenic line to study the role of ethylene production and perception in plant development during the low-temperature hardening (25/5 °C compared to 25/20 °C day/night), the subsequent severe chilling (5/5 °C day/night) and the following recovery period (25/20 °C day/night). Ethylene production was not different between the genotypes, but was approximately doubled through the hardening. Plants of the normal isogenic line immediately exposed to the severe chilling exhibited a greater chilling damage when compared to the non-hardened *Nr*-plants in terms of greater number of lesions, stronger reduction of shoot growth and lower chlorophyll fluorescence ratio (Fv/Fm), the latter also held true during the subsequent recovery. In contrast, the hardening of the normal plants induced an increased rate of leaf development and dry weight accumulation during the recovery period, whereas these hardening responses were suppressed and a reduction in lesion development was less pronounced in *Nr*-plants. These results demonstrated that ethylene production might trigger the development of chilling damage, while at the same time increase the tolerance of plants to subsequent chilling periods (Ciardi et al. 1997).

Ethylene biosynthesis is probably also involved in regulating of antifreeze activity in leaves of winter rye (*Secale cereale*), which arises from accumulation of antifreeze proteins in the apoplast (Yu et al. 2001). Endogenous ethylene production and antifreeze activity were detected within 12 and 48 h of exposure of plants to cold stress (5 °C) and also after 24 h of exposure to water deficit. Exposure of non-stressed (non-acclimated) rye plants to ethylene increased both antifreeze activity and the concentration of apoplastic protein in the apoplast. The important role of ethylene was further supported by the finding that the treatment with the ethylene-releasing compound, ethephon and ACC induced high levels of antifreeze activity at high temperatures, whereas these responses were blocked by application of SN.

5.5 Concluding Remarks

The role of ethylene in the stress response of plants is highly complex. The stress-mediated regulation of ethylene biosynthesis involves a co-ordinated activation particularly of ACS and ACO, which includes positive and/or negative feedback control loops and is based on the expression of genes of the ACS and ACO gene family (Fig. 5.1). Tissue specific, overlapping expression of several genes of the same family, each of which may be responsive to one or a set of stress factors, contributes to a physiological fine-tuning of the cell and consequently of the whole plant, which increases the flexibility of ethylene biosynthesis in a changing environment. Expression patterns among the *Arabidopsis* ACC gene family members even suggest (Tsuchisaka and Theologis 2004), that this fine-tuning involves the formation of heterodimeric isozymes ($_y\text{ACS}_z$ in Fig. 5.1). However, there is strong indication that ethylene biosynthesis is also highly dependent on post-transcriptional events. It further involves the action of other stress-sensitive plant hormones and of ROS, the latter seems to be important early mediators of different stresses (Fig. 5.1). Plants are even more adaptive by providing a flexible ethylene response; however, our current understanding of stress-mediated ethylene sensitivity is poor. It appears from many receptor studies that plants use the strategy of increased receptor synthesis as a dampening mechanism to slow down an ethylene response once it is initiated (Ciardi and Klee 2001). However, there is recent evidence that stress can also reduce receptor gene expression and indication that different receptor genes may respond to same stresses even in an opposite manner. Altogether, it can be assumed that plants use a differentiated and fine-tuned system on the ethylene perception side for adequate adjustment to the type, magnitude, duration and spatial distribution of the particular stress. An attempt was made in the present article to summarize the current knowledge of the involvement of ethylene production and ethylene sensitivity in plants exposed to physical and chemical stresses. Attention was given to the biochemistry and molecular biology of ethylene-related processes. However, the current knowledge of stress-mediated ethylene sensitivity is fragmentary so far and suggests that other plant hormones are substantially involved (Fig. 5.2). Future work is necessary to complete this picture.

Acknowledgements: This work was supported by the Ministry of Agriculture, Environmental Protection and Regional Planning of the State of Brandenburg, the Ministry for Agriculture, Nature Conservation and the Environment of the Free State of Thuringia, and the Federal Ministry for Consumer Protection, Food and Agriculture.

References

Abeles FB, Morgan PW, Salveit ME Jr (1992) Ethylene in plant biology, 2nd edn. Academic, New York

Alonso JM, Stepanova AN (2004) The ethylene signaling pathway. Science 306:1513–1515

Andersen L, Williams MH, Serek M (2004) Reduced water availability improves drought tolerance of potted miniature roses: is the ethylene pathway involved? J Hort Sci Biotech 79:1–13

Arimura G, Ozawa R, Nishioka T, Boland W, Koch T, Kühnemann F, Takabayashi J (2002) Herbivore-induced volatiles induce the emission of ethylene in neighboring lima bean plants. Plant J 29:87–98

Arteca JM, Arteca RN (1999) A multi-responsive gene encoding 1-aminocyclopropane-1-carboxylate synthase (ACS6) in mature *Arabidopsis* leaves. Plant Mol Biol 39:209–219

Balota M, Cristescu S, Payne WA, Hekkert STL, Laarhoven LJJ, Harren FJM (2004) Ethylene production of two wheat cultivars exposed to desiccation, heat, and paraquat-induced oxidation. Crop Sci 44:812–818

Banga M, Slaa EJ, Blom CWPM, Voesenek LACJ (1996a) Ethylene biosynthesis and accumulation under drained and submerged conditions – a comparative study of two *Rumex* species. Plant Physiol 112:229–237

Banga M, Blom CWPM, Voesenek LACJ (1996b) Sensitivity to ethylene: the key factor in submergence-induced shoot elongation of *Rumex*. Plant Cell Environ 19:1423–1430

Bar Y, Apelbaum A, Kafkafi U, Goren R (1998) Ethylene association with chloride stress in citrus plants. Sci Hort 73:99–109

Barry CS, Blume B, Bouzayen M, Cooper W, Hamilton AJ, Grierson D (1996) Differential expression of the 1-aminocyclopropane-1-carboxylate oxidase gene family of tomato. Plant J 9:525–535

Bassett CL, Artlip TS, Callahan AM (2002) Characterization of the peach homologue of the ethylene receptor, PpETR1, reveals some unusual features regarding transcript processing. Planta 215:679–688

Beltrano J, Montaldi E, Bartoli C, Carbone A (1997) Emission of water stress ethylene in wheat (*Triticum aestivum* L.) ears: effects of rewatering. Plant Growth Regul 21:121–126

Beltrano J, Ronco MG, Montaldi ER (1999) Drought stress syndrome in wheat is provoked by ethylene evolution imbalance and reversed by rewatering, aminoethoxyvinylglycine, or sodium benzoate. J Plant Growth Regul 18:59–64

Ben-Amor M, Flores B, Latche A, Bouzayen M, Pech JC, Romojaro F (1999) Inhibition of ethylene biosynthesis by antisense ACC oxidase RNA prevents chilling injury in Charentais cantaloupe melons. Plant Cell Environ 22:1579–1586

Blume B, Grierson D (1997) Expression of ACC oxidase promoter-GUS fusions in tomato and *Nicotiana plumbaginifolia* regulated by developmental and environmental stimuli. Plant J 12:731–746

Boller T (1991) Ethylene in pathogenesis and disease resistance. In: Mattoo AK, Suttle JS (eds) The plant hormone ethylene. CRC Press, Boca Raton, pp 293–314

Bors W, Langebartels C, Michel C, Sandermann H (1989) Polyamines as radical scavengers and protectants against ozone damage. Phytochemistry 28:1589–1595

Botella MA, del Amor F, Amoros A, Serrano M, Martinez V, Cerda A (2000) Polyamine, ethylene and other physico-chemical parameters in tomato (*Lycopersicon esculentum*) fruits as affected by salinity. Physiol Plant 109:428–434

Bouquin T, Lasserre E, Pradier J, Pech JC, Balague C (1997) Wound and ethylene induction of the ACC oxidase melon gene CM-ACO1 occurs via two direct and independent transduction pathways. Plant Mol Biol 35:1029–1035

Bradford KJ, Yang SF (1980) Xylem transport of 1-aminocyclopropane-1-carboxylic acid, an ethylene precursor, in waterlogged tomato plants. Plant Physiol 65:322–326

Cao WH, Dong Y, Zhang JS, Chen SY (2003) Characterization of an ethylene receptor homolog gene from rice. Science China Ser C Life Sci 46:370–378

Çelikel FG, Reid MS (2002) Storage temperature affects the quality of cut flowers from the Asteraceae. HortSci 37:148–150

Chen KM, Gong HJ, Chen GC, Zhang CL (2002) ACC and MACC biosynthesis and ethylene production in water-stressed spring wheat. Acta Bot Sinic 44:775–781

Ciardi J, Klee H (2001) Regulation of ethylene-mediated responses at the level of the receptor. Ann Bot 88:813–822

Ciardi JA, Deikman J, Orzolek MD (1997) Increased ethylene synthesis enhances chilling tolerance in tomato. Physiol Plant 101:333–340

Cohen E, Kende H (1986) The effect of submergence, ethylene and gibberellin on polyamines and their biosynthetic-enzymes in deep-water-rice internodes. Planta 169:498–504

Concellón A, Anon MC, Chaves AR (2005) Effect of chilling on ethylene production in eggplant fruit. Food Chem 92:63–69

Corbineau F, Berjak P, Pammenter N, Vinel D, Picard MA, Come D (2004) Reversible cellular and metabolic changes induced by dehydration in desiccation-tolerant wheat seedling shoots. Physiol Plant 122:28–38

Dat JF, Capelli N, Folzer H, Bourgeade P, Badot PM (2004) Sensing and signalling during plant flooding. Plant Physiol Biochem 42:273–282

Datta KS, Varma SK, Angrish R, Kumar B, Kumari P (1998) Alleviation of salt stress by plant growth regulators in *Triticum aestivum* L. Biol Plant 40:269–275

Davies WJ, Zhang JH (1991) Root signals and the regulation of growth and development of plants in drying Soil. Annu Rev Plant Physiol Plant Mol Biol 42:55–76

Drew MC, He CJ, Morgan PW (2000) Programmed cell death and aerenchyma formation in roots. Trends Plant Sci 5:123–127

Einset JW (1996) Differential expression of antisense in regenerated tobacco plants transformed with an antisense version of a tomato ACC oxidase gene. Plant Cell Tiss Organ Cult 46:137–141

El-Iklil Y, Karrou M, Benichou M (2000) Salt stress effect on epinasty in relation to ethylene production and water relations in tomato. Agron 20:399–406

El-Iklil Y, Karrou M, Mrabet R, Benichou M (2002) Salt stress effect on metabolite concentrations of *Lycopersicon esculentum* and *Lycopersicon sheesmanii*. Can J Plant Sci 82:177–183

El-Sharkawy I, Jones B, Li ZG, LeLièvre JM, Pech JC, Latche A (2003) Isolation and characterization of four ethylene perception elements and their expression during ripening in pears (*Pyrus communis* L.) with/without cold requirement. J Exp Bot 54:1615–1625

Else MA, Jackson MB (1998) Transport of 1-aminocyclopropane-1-carboxylic acid (ACC) in the transpiration stream of tomato (*Lycopersicon esculentum*) in relation to foliar ethylene production and petiole epinasty. Aust J Plant Physiol 25:453–458

English PJ, Lycett GW, Roberts JA, Jackson MB (1995) Increased 1-aminocyclopropane-1-carboxylic acid oxidase activity in shoots of flooded tomato plants raises ethylene production to physiologically active levels. Plant Physiol 109:1435–1440

Galil J (1968) An ancient technique for ripening Sycomore fruit in East-Mediterranean countries. Econ Bot 22:178

Gazzarrini S, McCourt P (2003) Cross-talk in plant hormone signalling: what *Arabidopsis* mutants are telling us. Ann Bot 91:605–612

Ge L, Liu JZ, Wong WS, Hsiao WLW, Chong K, Xu ZK, Yang SF, Kung SD, Li N (2000) Identification of a novel multiple environmental factor-responsive 1-aminocyclopropane-1-carboxylate synthase gene, NT-ACS2, from tobacco. Plant Cell Environ 23:1169–1182

Gelly M, Recasens I, Mata M, Arbones A, Rufat J, Girona J, Marsal J (2003) Effects of water deficit during stage II of peach fruit development and postharvest on fruit quality and ethylene production. J Hort Sci Biotech 78:324–330

Gerasopoulos D, Richardson DG (1997) Storage-temperature-dependent time separation of softening and chlorophyll loss from the autocatalytic ethylene pathway and other ripening events of 'Anjou' pears. J Am Soc Hort Sci 122:680–685

Gomez-Cadenas A, Tadeo FR, Talon M, Primo-Millo E (1996) Leaf abscission induced by ethylene in water-stressed intact seedlings of Cleopatra mandarin requires previous abscisic acid accumulation in roots. Plant Physiol 112:401–408

Gomez-Cadenas A, Tadeo FR, Primo-Millo E, Talon M (1998) Involvement of abscisic acid and ethylene in the responses of citrus seedlings to salt shock. Physiol Plant 103:475–484

Gomez-Cadenas A, Arbona V, Jacas J, Primo-Millo E, Talon M (2002) Abscisic acid reduces leaf abscission and increases salt tolerance in citrus plants. J Plant Growth Regul 21:234–240

Grichko VP, Glick BR (2001) Flooding tolerance of transgenic tomato plants expressing the bacterial enzyme ACC deaminase controlled by the 35S, rolD or PRB-1b promoter. Plant Physiol Biochem 39:19–25

Guo HW, Ecker JR (2004) The ethylene signaling pathway: new insights. Curr Opin Plant Biol 7:40–49

He CJ, Finlayson SA, Drew MC, Jordan WR, Morgan PW (1996) Ethylene biosynthesis during aerenchyma formation in roots of maize subjected to mechanical impedance and hypoxia. Plant Physiol 112:1679–1685

He XJ, Zhang ZG, Yan DQ, Zhang JS, Chen SY (2004) A salt-responsive receptor-like kinase gene regulated by the ethylene signaling pathway encodes a plasma membrane serine/threonine kinase. Theor Appl Gen 109:377–383

Howe GA (2004) Jasmonates as signals in the wound response. J Plant Growth Regul 23:223–237

Hsu SY, Hsu YT, Kao CH (2003) Ammonium ion, ethylene, and abscisic acid in polyethylene glycol-treated rice leaves. Biol Plant 46:239–242

Hu WW, Gong HB, Pua EC (2005) Identification of stress-induced mitochondrial proteins in cultured tobacco cells. Physiol Plant 124:25–40

Hussain A, Black CR, Taylor LB, Roberts JA (1999) Soil compaction. A role for ethylene in regulating leaf expansion and shoot growth in tomato? Plant Physiol 121:1227–1237

Hussain A, Black CR, Taylor IB, Roberts JA (2000) Does an antagonistic relationship between ABA and ethylene mediate shoot growth when tomato (*Lycopersicon esculentum* Mill.) plants encounter compacted soil? Plant Cell Environ 23:1217–1226

Hyodo H (1991) Stress/wound ethylene. In: Mattoo AK, Suttle JS (eds) The plant hormone ethylene. CRC Press, Boca Raton, pp 43–64

Ichimura K, Suto K (1998) Role of ethylene in acceleration of flower senescence by filament wounding in *Portulaca* hybrid. Physiol Plant 104:603–607

Jackson MB (1985) Ethylene and responses of plants to soil waterlogging and submergence. Annu Rev Plant Physiol Plant Mol Biol 36:145–174

Jackson MB, Armstrong W (1999) Formation of aerenchyma and the processes of plant ventilation in relation to soil flooding and submergence. Plant Biol 1:274–287

Jackson MB, Campbell DJ (1976) Waterlogging and petiole epinasty in tomato – role of ethylene and low oxygen. New Phytol 76:21–29

Jaffe MJ, Forbes S (1993) Thigmomorphogenesis – the effect of mechanical perturbation on plants. Plant Growth Regul 12:313–324

Johnson KA, Sistrunk ML, Polisensky DH, Braam J (1998) *Arabidopsis thaliana* responses to mechanical stimulation do not require ETR1 or EIN2. Plant Physiol 116:643–649

Jones ML, Woodson WR (1999) Differential expression of three members of the 1-aminocyclopropane-1-carboxylate synthase gene family in carnation. Plant Physiol 119:755–764

Kadner R, Druege U (2004) Role of ethylene action in ethylene production and poststorage leaf senescence and survival of pelargonium cuttings. Plant Growth Regul 43:187–196

Kadner R, Druege U, Kühnemann F (2000) Ethylene emission of cuttings of *Pelargonium* during the storage at different temperatures. Gartenbauwiss 65:272–279

Kalantari KM, Smith AR, Hall MA (2000) The effect of water stress on 1-(malonylamino)cyclopropane-1-carboxylic acid concentration in plant tissues. Plant Growth Regul 31:183–193

Kangasjärvi J, Talvinen J, Utriainen M, Karjalainen R (1994) Plant defense systems induced by ozone. Plant Cell Environ 17:783–794

Kanna M, Tamaoki M, Kubo A, Nakajima N, Rakwal R, Agrawal GK, Tamogami S, Ioki M, Ogawa D, Saji H, Aono M (2003) Isolation of an ozone-sensitive and jasmonate-semi-insensitive *Arabidopsis* mutant (oji1). Plant Cell Physiol 44:1301–1310

Kato M, Hayakawa Y, Hyodo H, Ikoma Y, Yano M (2000) Wound-induced ethylene synthesis and expression and formation of 1-aminocyclopropane-1-carboxylate (ACC) synthase, ACC oxidase, phenylalanine ammonia-lyase, and peroxidase in wounded mesocarp tissue of *Cucurbita maxima*. Plant Cell Physiol 41:440–447

Katz E, Riov J, Weiss D, Goldschmidt EE (2005) The climacteric-like behaviour of young, mature and wounded citrus leaves. J Exp Bot 56:1359–1367

Ke DS, Sun GC (2004) The effect of reactive oxygen species on ethylene production induced by osmotic stress in etiolated mungbean seedling. Plant Growth Regul 44:199–206

Kende H, van der Knaap E, Cho HT (1998) Deepwater rice: a model plant to study stem elongation. Plant Physiol 118:1105–1110

Ketsa S, Chidtragool S, Klein JD, Lurie S (1999) Ethylene synthesis in mango fruit following heat treatment. Postharvest Biol Technol 15:65–72

Klee HJ (2004) Ethylene signal transduction. Moving beyond *Arabidopsis*. Plant Physiol 135:660–667

Lara I, Vendrell M (2003) Cold-induced ethylene biosynthesis is differentially regulated in peel and pulp tissues of 'Granny Smith' apple fruit. Postharvest Biol Technol 29:109–119

Larcher W (1995) Physiological plant ecology – ecophysiology and stress physiology of functional groups, 3rd edn. Springer, Berlin Heidelberg New York

Larrigaudiere C, Graell J, Salas J, Vendrell M (1997) Cultivar differences in the influence of a short period of cold storage on ethylene biosynthesis in apples. Postharvest Biol Technol 10:21–27

LeLièvre JM, Tichit L, Dao P, Fillion L, Nam YW, Pech JC, Latche A (1997) Effects of chilling on the expression of ethylene biosynthetic genes in Passe-Crassane pear (*Pyrus communis* L) fruits. Plant Mol Biol 33:847–855

Levitt J (1980) Responses of plants to environmental stresses. Academic, New York

Li CZ, Wang GX (2004) Interactions between reactive oxygen species, ethylene and polyamines in leaves of *Glycyrrhiza inflata* seedlings under root osmotic stress. Plant Growth Regul 42:55–60

Li CZ, Jiao J, Wang GX (2004) The important roles of reactive oxygen species in the relationship between ethylene and polyamines in leaves of spring wheat seedlings under root osmotic stress. Plant Sci 166:303–315

Lin JN, Kao CH (1998) Water stress, ammonium, and leaf senescence in detached rice leaves. Plant Growth Regul 25:165–169

Liu JH, LeeTamon SH, Reid DM (1997) Differential and wound-inducible expression of 1-aminocyclopropane-1-carboxylate oxidase genes in sunflower seedlings. Plant Mol Biol 34:923–933

Liu FL, VanToai T, Moy LP, Bock G, Linford LD, Quackenbush J (2005) Global transcription profiling reveals comprehensive insights into hypoxic response in *Arabidopsis*. Plant Physiol 137:1115–1129

Lutts S, Kinet JM, Bouharmont J (1996) Ethylene production by leaves of rice (*Oryza sativa* L) in relation to salinity tolerance and exogenous putrescine application. Plant Sci 116:15–25

Ma QH, Wang XM (2003) Characterization of an ethylene receptor homologue from wheat and its expression during leaf senescence. J Exp Bot 54:1489–1490

Martínez-Romero D, Serrano M, Valero D (2003) Physiological changes in pepino (*Solanum muricatum* Ait.) fruit stored at chilling and non-chilling temperatures. Postharvest Biol Technol 30:177–186

Mathooko FM, Tsunashima Y, Owino WZO, Kubo Y, Inaba A (2001) Regulation of genes encoding ethylene biosynthetic enzymes in peach (*Prunus persica* L.) fruit by carbon dioxide and 1-methylcyclopropene. Postharvest Biol Technol 21:265–281

Mekhedov SL, Kende H (1996) Submergence enhances expression of a gene encoding 1-aminocyclopropane-1-carboxylate oxidase in deepwater rice. Plant Cell Physiol 37:531–537

Métraux JP, Kende H (1983) The role of ethylene in the growth-response of submerged deepwater rice. Plant Physiol 72:441–446

Michaeli R, Philosoph-Hadas S, Riov J, Meir S (1999a) Chilling-induced leaf abscission of *Ixora coccinea* plants. I. Induction by oxidative stress via increased sensitivity to ethylene. Physiol Plant 107:166–173

Michaeli R, Riov J, Philosoph-Hadas S, Meir S (1999b) Chilling-induced leaf abscission of *Ixora coccinea* plants. II. Alteration of auxin economy by oxidative stress. Physiol Plant 107:174–180

Moeder W, Barry CS, Tauriainen AA, Betz C, Tuomainen J, Utriainen M, Grierson D, Sandermann H, Langebartels C, Kangasjärvi J (2002) Ethylene synthesis regulated by biphasic induction of 1-aminocyclopropane-1-carboxylic acid synthase and 1-aminocyclopropane-1-carboxylic acid oxidase genes is required for hydrogen peroxide accumulation and cell death in ozone-exposed tomato. Plant Physiol 130:1918–1926

Morgan PW, Drew MC (1997) Ethylene and plant responses to stress. Physiol Plant 100:620–630

Mudd JB (1996) Biochemical basis for the toxicity of ozone. In: Iqbal M, Yunus M (eds) Plant response to air pollution. Wiley, New York, pp 267–283

Munns R (1993) Physiological processes limiting plant-growth in saline soils – some dogmas and hypotheses. Plant Cell Environ 16:15–24

Nakajima N, Matsuyama T, Tamaoki M, Saji H, Aono M, Kubo A, Kondo N (2001) Effects of ozone exposure on the gene expression of ethylene biosynthetic enzymes in tomato leaves. Plant Physiol Biochem 39:993–998

Nakajima N, Itoh T, Takikawa S, Asai N, Tamaoki M, Aono M, Kubo A, Azumi Y, Kamada H, Saji H (2002) Improvement in ozone tolerance of tobacco plants with an antisense DNA for 1-aminocyclopropane-1-carboxylate synthase. Plant Cell Environ 25:727–735

Nakano R, Inoue S, Kubo Y, Inaba A (2002) Water stress-induced ethylene in the calyx triggers autocatalytic ethylene production and fruit softening in 'Tonewase' persimmon grown in a heated plastic-house. Postharvest Biol Technol 25:293–300

Nie XZ, Singh RP, Tai GCC (2002) Molecular characterization and expression analysis of 1-aminocyclopropane-1-carboxylate oxidase homologs from potato under abiotic and biotic stresses. Genome 45:905–913

O'Donnell PJ, Calvert C, Atzorn R, Wasternack C, Leyser HMO, Bowles DJ (1996) Ethylene as a signal mediating the wound response of tomato plants. Science 274:1914–1917

Ouvrard O, Cellier F, Ferrare K, Tousch D, Lamaze T, Dupuis JM, CasseDelbart F (1996) Identification and expression of water stress- and abscisic acid-regulated genes in a drought-tolerant sunflower genotype. Plant Mol Biol 31:819–829

Overmyer K, Tuominen H, Kettunen R, Betz C, Langebartels C, Sandermann H, Kangasjärvi J (2000) Ozone-sensitive *Arabidopsis* rcd1 mutant reveals opposite roles for ethylene and jasmonate signaling pathways in regulating superoxide-dependent cell death. Plant Cell 12:1849–1862

Owino WO, Nakano R, Kubo Y, Inaba A (2002) Differential regulation of genes encoding ethylene biosynthesis enzymes and ethylene response sensor ortholog during ripening and in response to wounding in avocados. J Am Soc Hort Sci 127:520–527

Pech JC, Latché A, Bouzayen M (2004) Ethylene biosynthesis. In: Davies PJ (ed) Plant hormones: biosynthesis, signal transduction, action, 3rd edn. Kluwer, Dordrecht, pp 115–136

Peeters AJM, Cox MCH, Benschop JJ, Vreeburg RAM, Bou J, Voesenek LACJ (2002) Submergence research using *Rumex palustris* as a model; looking back and going forward. J Exp Bot 53:391–398

Pell EJ, Schlagnhaufer CD, Arteca RN (1997) Ozone-induced oxidative stress: mechanisms of action and reaction. Physiol Plant 100:264–273

Pereira-Netto AB, McCown BH (1999) Thermally induced changes in shoot morphology of *Hancornia speciosa* microcultures: evidence of mediation by ethylene. Tree Physiol 19:733–740

Purvis AC, Shewfelt RL, Gegogeine JW (1995) Superoxide production by mitochondria isolated from green bell pepper fruit. Physiol Plant 94:743–749

Rao MV, Lee H, Creelman RA, Mullet JE, Davis KR (2000) Jasmonic acid signaling modulates ozone-induced hypersensitive cell death. Plant Cell 12:1633–1646

Rao MV, Lee H, Davis KR (2002) Ozone-induced ethylene production is dependent on salicylic acid, and both salicylic acid and ethylene act in concert to regulate ozone-induced cell death. Plant J 32:447–456

Rieu I, Cristescu SM, Harren FJM, Huibers W, Voesenek LACJ, Mariani C, Vriezen WH (2005) RP-ACS1, a flooding-induced 1-aminocyclopropane-1-carboxylate synthase gene of *Rumex palustris*, is involved in rhythmic ethylene production. J Exp Bot 56:841–849

Salveit ME, Morris LL (1990) Overview on chilling injury of horticultural crops. In: Wang CY (ed) Chilling injury of horticultural crops. CRS Press, Boca Raton, pp 3–15

Schlagnhaufer CD, Arteca RN, Pell EJ (1997) Sequential expression of two 1-aminocyclopropane-1-carboxylate synthase genes in response to biotic and abiotic stresses in potato (*Solanum tuberosum* L.) leaves. Plant Mol Biol 35:683–688

Schraudner M, Moeder W, Wiese C, van Camp W, Inze D, Langebartels C, Sandermann H (1998) Ozone-induced oxidative burst in the ozone biomonitor plant, tobacco Bel W3. Plant J 16:235–245

Sharp RE (2002) Interaction with ethylene: changing views on the role of abscisic acid in root and shoot growth responses to water stress. Plant Cell Environ 25:211–222

Sharp RE, LeNoble ME (2002) ABA, ethylene and the control of shoot and root growth under water stress. J Exp Bot 53:33–37

Shiomi S, Yamamoto M, Ono T, Kakiuchi K, Nakamoto J, Nakatsuka A, Kubo Y, Nakamura R, Inaba A, Imaseki H (1998) cDNA cloning of ACC synthase and ACC oxidase genes in cucumber fruit and their differential expression by wounding and auxin. J Jpn Soc Hort Sci 67:685–692

Shiomi S, Yamamoto M, Nakamura R, Inaba A (1999) Expression of ACC synthase and ACC oxidase genes in melons harvested at different stages of maturity. J Jpn Soc Hort Sci 68:10–17

Shiu OY, Oetiker JH, Yip WK, Yang SF (1998) The promoter of LE-ACS7, an early flooding-induced 1-aminocyclopropane-1-carboxylate synthase gene of the tomato, is tagged by a Sol3 transposon. Proc Natl Acad Sci USA 95:10334–10339

Sobeih WY, Dodd IC, Bacon MA, Grierson D, Davies WJ (2004) Long-distance signals regulating stomatal conductance and leaf growth in tomato (*Lycopersicon esculentum*) plants subjected to partial root-zone drying. J Exp Bot 55:2353–2363

Song LL, Peng YH (2004) Effect of cold storage on sensitivity of cut lily to ethylene. J Hort Sci Biotech 79:723–728

Sunohara Y, Ikeda S, Murata Y, Sakurai N, Noma Y (2002) Effects of trampling on morphology and ethylene production in asiatic plantain. Weed Sci 50:479–484

Suzuki H, Tsuruhara A, Tezuka T (2001) Regulations of the C_2H_4-forming system and the H_2O_2-scavenging system by heat treatment associated with self incompatibility in lily. Sex Plant Reprod 13:201–208

Tamaoki M, Matsuyama T, Kanna M, Nakajima N, Kubo A, Aono M, Saji H (2003) Differential ozone sensitivity among *Arabidopsis* accessions and its relevance to ethylene synthesis. Planta 216:552–560

Tardieu F, Zhang J, Katerji N, Bethenod O, Palmer S, Davies WJ (1992) Xylem ABA controls the stomatal conductance of field-grown maize subjected to soil compaction or soil drying. Plant Cell Environ 15:193–197

Tatsuki M, Mori H (1999) Rapid and transient expression of 1-aminocyclopropane-1-carboxylate synthase isogenes by touch and wound stimuli in tomato. Plant Cell Physiol 40:709–715

Tatsuki M, Mori H (2001) Phosphorylation of tomato 1-aminocyclopropane-1-carboxylic acid synthase, LE-ACS2, at the C-terminal region. J Biol Chem 276:28051–28057

Tingey DT, Standley C, Field RW (1976) Stress ethylene evolution – measure of ozone effects on plants. Atmos Environ 10:969–974

Treshow M, Anderson FK (1989) Plant stress from air pollution. Wiley, New York

Tsuchisaka A, Theologis A (2004) Unique and overlapping expression patterns among the *Arabidopsis* 1-amino-cyclopropane-1-carboxylate synthase gene family members. Plant Physiol 136:2982–3000

Tudela D, Primo-Millo E (1992) 1-aminocyclopropane-1-carboxylic acid transported from roots to shoots promotes leaf abscission in Cleopatra mandarine (*Citrus-reshni* Hort. Ex Tan.) seedlings rehydrated after water-stress. Plant Physiol 100:131–137

Tuomainen J, Betz C, Kangasjärvi J, Ernst D, Yin ZH, Langebartels C, Sandermann H (1997) Ozone induction of ethylene emission in tomato plants: regulation by differential accumulation of transcripts for the biosynthetic enzymes. Plant J 12:1151–1162

Vahala J, Schlagnhaufer CD, Pell EJ (1998) Induction of an ACC synthase cDNA by ozone in light-grown *Arabidopsis thaliana* leaves. Physiol Plant 103:45–50

Vahala J, Ruonala R, Keinanen M, Tuominen H, Kangasjärvi J (2003) Ethylene insensitivity modulates ozone-induced cell death in birch. Plant Physiol 132:185–195

Verlinden S, Woodson WR (1998) The physiological and molecular responses of carnation flowers to high temperature. Postharvest Biol Technol 14:185–192

Visser EJW, Cohen JD, Barendse GWM, Blom CWPM, Voesenek LACJ (1996) An ethylene-mediated increase in sensitivity to auxin induces adventitious root formation in flooded *Rumex palustris* Sm. Plant Physiol 112:1687–1692

Voesenek LACJ, Vriezen WH, Smekens MJE, Huitink FHM, Bogemann GM, Blom CWPM (1997) Ethylene sensitivity and response sensor expression in petioles of *Rumex* species at low O_2 and high CO_2 concentrations. Plant Physiol 114:1501–1509

Vriezen WH, vanRijn CPE, Voesenek LACJ, Mariani C (1997) A homolog of the *Arabidopsis thaliana* ERS gene is actively regulated in *Rumex palustris* upon flooding. Plant J 11:1265–1271

Vriezen WH, Hulzink R, Mariani C, Voesenek LACJ (1999) 1-aminocyclopropane-1-carboxylate oxidase activity limits ethylene biosynthesis in *Rumex palustris* during submergence. Plant Physiol 121:189–195

Wang H, Huang ZJ, Chen Q, Zhang ZJ, Zhang HB, Wu YM, Huang DF, Huang RF (2004) Ectopic overexpression of tomato JERF3 in tobacco activates downstream gene expression and enhances salt tolerance. Plant Mol Biol 55:183–192

Wang NN, Shih MC, Li N (2005) The GUS reporter-aided analysis of the promoter activities of *Arabidopsis* ACC synthase genes AtACS4, AtACS5, and AtACS7 induced by hormones and stresses. J Exp Bot 56:909–920

Wang TW, Arteca RN (1992) Effects of low O_2 root stress on ethylene biosynthesis in tomato plants (*Lycopersicon esculentum* Mill. cv Heinz 1350). Plant Physiol 98:97–100

Watanabe H, Saigusa M, Hase S, Hayakawa T, Satoh S (2004) Cloning of a cDNA encoding an ETR2-like protein (Os-ERL1) from deep water rice (*Oryza sativa* L.) and increase in its mRNA level by submergence, ethylene, and gibberellin treatments. J Exp Bot 55:1145–1148

Watanabe T, Sakai S (1998) Effects of active oxygen species and methyl jasmonate on expression of the gene for a wound-inducible 1-aminocyclopropane-1-carboxylate synthase in winter squash (*Cucurbita maxima*). Planta 206:570–576

Watanabe T, Seo S, Sakai S (2001) Wound-induced expression of a gene for 1-aminocyclopropane-1-carboxylate synthase and ethylene production are regulated by both reactive oxygen species and jasmonic acid in *Cucurbita maxima*. Plant Physiol Biochem 39:121–127

Wi SJ, Park KY (2002) Antisense expression of carnation cDNA encoding ACC synthase or ACC oxidase enhances polyamine content and abiotic stress tolerance in transgenic tobacco plants. Mol Cells 13:209–220

Wong WS, Ning W, Xu PL, Kung SD, Yang SF, Li N (1999) Identification of two chilling-regulated 1-aminocyclopropane-1-carboxylate synthase genes from citrus (*Citrus sinensis* Osbeck) fruit. Plant Mol Biol 41:587–600

Wong WS, Li GG, Ning W, Xu ZF, Hsiao WLW, Zhang LY, Li N (2001) Repression of chilling-induced ACC accumulation in transgenic citrus by over-production of antisense 1-aminocyclopropane-1-carboxylate synthase RNA. Plant Sci 161:969–977

Xie C, Zhang ZG, Zhang JS, He XJ, Cao WH, He SJ, Chen SY (2002) Spatial expression and characterization of a putative ethylene receptor protein NTHK1 in tobacco. Plant Cell Physiol 43:810–815

Yu SJ, Kim S, Lee JS, Lee DH (1998) Differential accumulation of transcripts for ACC synthase and ACC oxidase homologs in etiolated mung bean hypocotyls in response to various stimuli. Mol Cells 8:350–358

Yu XM, Griffith M, Wiseman SB (2001) Ethylene induces antifreeze activity in winter rye leaves. Plant Physiol 126:1232–1240

Zapata PJ, Serrano M, Pretel MT, Amoros A, Botella A (2003) Changes in ethylene evolution and polyamine profiles of seedlings of nine cultivars of *Lactuca sativa* L. in response to salt stress during germination. Plant Sci 164:557–563

Zapata PJ, Serrano M, Pretel AT, Amoros A, Botella MA (2004) Polyamines and ethylene changes during germination of different plant species under salinity. Plant Sci 167:781–788

Zarembinski TI, Theologis A (1993) Anaerobiosis and plant-growth hormones induce 2 genes encoding 1-aminocyclopropane-1-carboxylate synthase in rice (*Oryza sativa* L). Mol Biol Cell 4:363–373

Zarembinski TI, Theologis A (1997) Expression characteristics of OS-ACS1 and OS-ACS2, two members of the 1-aminocyclopropane-1-carboxylate synthase gene family in rice (*Oryza sativa* L cv Habiganj Aman II) during partial submergence. Plant Mol Biol 33:71–77

Zeroni M, Benyehos S, Galil J (1972) Relationship between ethylene and growth of *Ficus sycomorus*. Plant Physiol 50:378

Zhang JS, Xie C, Shen YG, Chen SY (2001) A two-component gene (NTHK1) encoding a putative ethylene-receptor homolog is both developmentally and stress regulated in tobacco. Theor Appl Gen 102:815–824

Zhao XC, Schaller GE (2004) Effect of salt and osmotic stress upon expression of the ethylene receptor ETR1 in *Arabidopsis thaliana*. FEBS Lett 562:189–192

Zhou ZY, Vriezen W, van Caeneghem W, van Montagu M, van der Straeten D (2001) Rapid induction of a novel ACC synthase gene in deepwater rice seedlings upon complete submergence. Euphytica 121:137–143

6 Ethylene in the *Rhizobium*-Legume Symbiosis

Jeroen Den Herder, Sofie Goormachtig, Marcelle Holsters

6.1 An Introduction to Legume Nodulation

The plant hormone ethylene plays an important role in plant-bacterium interactions. During pathogen attack, ethylene mediates defense responses to reduce or enhance disease symptoms depending on the pathogenic strategy (Hoffman et al. 1999). Ethylene also has a major function in the beneficial rhizobial and arbuscular mycorrhizal symbioses (Guinel and Geil 2002). The *Rhizobium*-legume interaction results in the formation of new organs, the nodules, on roots of compatible host plants. Inside the central nodule cells, bacteria are housed as symbiosomes, which are horizontally acquired organelles that enzymatically reduce atmospheric dinitrogen to provide their host with ammonia. In agriculture, symbiotic nitrogen fixation can be an environmentally friendly alternative for nitrate fertilization, which pollutes ground waters because of excess application. The capacity to establish a nitrogen-fixing symbiosis with leguminous plants was first thought to be restricted to α-proteobacteria of the genera *Rhizobium*, *Bradyrhizobium*, *Sinorhizobium*, *Mesorhizobium*, *Allorhizobium*, and *Azorhizobium*. However, bacteria from outside the Rhizobiaceae family can also associate with legumes, such as strains of *Methylobacterium*, but also members of the β-proteobacteria, such as some *Burkholderia* sp. strains and *Ralstonia taiwanensis* (Moulin et al. 2001; Sy et al. 2001; Chen et al. 2003). Hence, nowadays the term 'rhizobia' is used to designate all soil bacteria capable of establishing a nitrogen-fixing symbiosis with legume roots.

The *Rhizobium*-legume interaction starts when root exudates, such as flavonoids, trigger the transcription of nodulation (*nod*) genes in compatible bacteria, resulting in the production of nodulation factors (NFs), which act as return signals to initiate nodule development. NFs are lipochito-oligosaccharides that consist of a backbone of *N*-acetylglucosamine residues with an acyl chain at the non-reducing end and strain-specific modifications, such as methylation, acetylation, carbamoylation, arabinosylation, and fucosylation at both reducing and non-reducing termini (D'Haeze and Holsters 2002; Geurts and Bisseling 2002). NF perception by

---

Department of Plant Systems Biology, Flanders Interuniversity Institute for Biotechnology (VIB), Ghent University, Technologiepark 927, 9052 Gent, Belgium

Ethylene Action in Plants
(ed. by N.A. Khan)
© Springer-Verlag Berlin Heidelberg 2006

specific receptors in compatible host plants activates a genetic program for organ development concomitant with bacterial invasion.

The most common and best-characterized *Rhizobium*-legume interactions start via intracellular invasion of root hairs. The root hair curling (RHC) nodulation mechanism occurs in major crops (e.g., pea, common bean, soybean, vetch, and alfalfa) as well as in the model legumes *Medicago truncatula* and *Lotus japonicus*. Bacteria colonize developing root hairs located above the root meristem (zone I root hairs) and produce NFs that interfere with the root hair growth, causing the formation of a three-dimensional curl that entraps a bacterial microcolony (Kijne 1992). After local cell wall hydrolysis, the plant plasma membrane is invaginated to form an infection thread (IT) that guides dividing bacteria to the base of the root hair. Simultaneously, NF signaling triggers reinitiation of the cell cycle in cortical cells to create a nodule primordium in the inner cortex. In the outer cortex, the cell cycle is initiated but arrested before mitosis, resulting in the formation of cytoplasmic bridges, also called pre-infection threads (PITs), through which the ITs progress. Transcellular ITs fuse with the distal cell wall and proceed from cell to cell in a repetitive process of membrane invagination, tip growth, and cell wall fusion (Gage and Margolin 2000; Gage 2004). When the inward-growing IT meets the outward-growing nodule primordium, the bacteria are engulfed by the plant membrane and internalized in the plant cell to differentiate into nitrogen-fixing bacteroids. For an extensive description of these processes, we refer to Brewin (2004).

In an alternative way for nodule initiation that differs from RHC nodulation in the primary invasion stages, infection does not require root hairs but occurs intercellularly at lateral root bases (LRBs), where a fissure in the epidermis exposes cortex cells to the environment. LRB nodulation by crack-entry invasion has been best studied in the semi-aquatic tropical legume *Sesbania rostrata*. This legume has versatile growth and nodulation features as an adaptation to temporarily flooded habitats. Upon inoculation of well-aerated roots with a compatible microsymbiont, such as *Azorhizobium caulinodans*, RHC nodulation occurs in zone I (Goormachtig et al. 2004b). However, upon flooding and at positions of stem-located adventitious rootlets, nodules form via crack-entry invasion. Bacteria colonize the fissure that is present at the base of the lateral roots or the adventitious rootlets. Bacterial NFs trigger local cell death of a few cortical cells to create intercellular spaces that are colonized by the rhizobia to form infection pockets (IPs) (D'Haeze et al. 2003). As during RHC invasion, a nodule primordium is initiated in the mid-inner cortex. From the IPs, the bacteria migrate via inter- and intracellular ITs toward the developing primordium, in which they are again internalized to form symbiosomes (Ndoye et al. 1994; Goormachtig et al. 1998).

Legume nodules can be of several types (Hirsch 1992). Most legume species, including *M. truncatula*, develop indeterminate nodules characterized by an elongated shape that is caused by the presence of a persistent apical meristem. In these indeterminate nodules, several zones can be distinguished: the distal meristem that delivers new cells to the infection zone where bacteria are internalized, an interzone with amyloplast accumulation and differentiation of bacteroids, a fixation zone with plant cells harboring N_2-fixing symbiosomes interspersed with non-infected cells, and a senescent zone (Vasse et al. 1990; Pawlowski and Bisseling 1996; Timmers et al. 1999). In *L. japonicus* and in a number of tropical legumes, nodules are of the determinate type with a typical round shape that results from the cessation of meristem activity after nodule initiation and growth of the nodule mainly by cell expansion (Sprent 2002). Interestingly, in *S. rostrata*, both types of nodules can be found dependent on the growth conditions. Round determinate nodules are present on hydroponic roots while indeterminate cylindrical nodules with a persistent meristem occur on aerated roots (Fernández-López et al. 1998). The factors that control the switch will be discussed below.

Many plant hormones are involved in the nodulation process. The balance between auxin and cytokinin presumably plays a role in setting the landscape for nodule development (Ferguson and Mathesius 2003). Cytokinins were detected in *Pisum sativum* (pea) nodules (Syōno and Torrey 1976). N-(1-naphthyl)phthalamic acid (NPA), an inhibitor of auxin transport, induced pseudonodules on *Medicago sativa* (alfalfa) roots that contained transcripts for the early nodulin gene *ENOD2* (Hirsch et al. 1989). Likewise, cytokinin application triggered *ENOD2* expression in *S. rostrata* roots and *ENOD40* induction in alfalfa and *Trifolium repens* (white clover) (Dehio and de Bruijn 1992; Fang and Hirsch 1998; Mathesius et al. 2000). Auxin transport inhibition preceded nodule formation in roots of white clover and the expression of an auxin-responsive reporter construct *GH3:gusA* was rapidly, but transiently, down-regulated after inoculation, followed by an up-regulation at the site of nodule initiation (Mathesius et al. 1998). Changes in endogenous hormone levels may be the consequence of NF perception in the legume host (Hirsch and Fang 1994). Recently, also gibberellic acid (GA) has been found to be involved in nodule initiation and development (Ferguson et al. 2005). In *S. rostrata*, an enzyme of the GA biosynthesis pathway, a gibberellin-20-oxidase, is produced during LRB nodulation, whereas chlormequat chloride, an inhibitor of gibberellin synthesis, blocks LRB nodulation when applied prior to infection (Lievens et al. 2005).

In this chapter we will discuss the involvement of ethylene at different stages of the nodulation process. We will consider both nodulation via RHC in *M. truncatula* and other legumes that are sensitive to ethylene, and LRB nodulation with crack-entry invasion in *S. rostrata*, which depends on ethylene.

6.2 Ethylene-Sensitive RHC Nodulation

6.2.1 Pharmacological Evidence

Plenty of data are available regarding pharmacological experiments that involve ethylene, ethylene-releasing molecules, and inhibitors of ethylene synthesis and perception in different legume species. More than 30 years ago, Grobbelaar et al. (1971) observed that nodulation of *Phaseolus vulgaris* (common bean) was inhibited when ethylene was applied to the roots. Exogenous ethylene also decreased the number of nodules in pea. Not the number of primary infections was reduced, but fewer bacteria could enter the outer cortical cells (Lee and LaRue 1992). Moreover, nodulation of the pea mutant E107 (*brz*) that had a low number of infections of which only a small percentage passed the epidermis, was partly restored by L-α-(2-aminoethoxyvinyl) glycine (AVG) or silver ions, inhibitors of ethylene synthesis and perception, respectively. Instead of increasing the number of initial infections, silver treatment allowed more ITs to pass the epidermis and proceed toward the cortex (Guinel and LaRue 1992).

Also in *M. sativa*, treatment with AVG increased the number of nodules (Peters and Crist-Estes 1989). Application of 1-aminocyclopropane-1-carboxylic acid (ACC), the direct precursor of ethylene, to *M. truncatula* roots had an inhibitory effect on the nodule number (Penmetsa and Cook 1997). When ACC was added during initiation of infection, i.e., before the first nodule primordia were evident, nodulation was blocked. After nodule primordia were visible macroscopically, further nodule development was not affected by ACC, suggesting that sustained rhizobial infection may have acquired insensitivity to ethylene (Penmetsa and Cook 1997). Additionally, Oldroyd et al. (2001) demonstrated that in *M. truncatula* the infection frequency was influenced by exogenous ACC or AVG. The plants were grown on plates with increasing levels of ACC or AVG and the number of infection events were counted, including curled root hairs with a bacterial microcolony, in which no ITs were visible. In this experimental set-up, the number of infection events increased with decreasing levels of ethylene (Oldroyd et al. 2001).

In *L. japonicus* as well as in *Macroptilium atropurpureum* (siratro), two species that form determinate nodules, ACC reduced the number of nodules whereas AVG and silver ions enhanced nodulation (Nukui et al. 2000; Yuhashi et al. 2000). The formation of primordia and nodules was enhanced only at later stages after application of silver ions, while this increase started earlier after addition of AVG, suggesting different effects of altered ethylene synthesis and perception (Nukui et al. 2004).

Vicia sativa subsp. *nigra* (vetch) developed a thick short root (Tsr) phenotype upon growth in the light and subsequent inoculation with its bacterial partner. The roots were twice as thick as normal and had a reduced length and an increased number of root hairs. Nodulation was delayed and nodules were formed at sites of lateral root emergence rather than on the main root. After addition of AVG, the development of the phenotype was suppressed; the plants nodulated earlier and the nodules were located on the primary root (Zaat et al. 1989). Later, the Tsr phenotype was characterized by a swelling of the cortical cells, which corresponded with a reorientation of the microtubules from a transverse to a longitudinal direction, with cell wall modifications and frequent absence of middle lamellae (van Spronsen et al. 1995). Similar changes could be induced by the ethylene-releasing molecule 2-chloroethylphosphonic acid (ethephon) and were inhibited by AVG. An excess of ethylene production, triggered by NFs when the roots are exposed to light, might cause the Tsr phenotype. The ethylene-related changes in the cortex would inhibit nodulation probably by preventing formation of PITs and by reducing formation of nodule primordia (van Spronsen et al. 1995). Consistent with this hypothesis was the observation that after growth in the light AVG restored the Tsr phenotype of *L. japonicus*, in which PITs were found, while it did not restore the delayed nodulation in bean where no cytoplasmic bridges are formed (van Spronsen et al. 2001).

Interestingly, not all plant species have features of ethylene-inhibited nodulation. The best-studied example of non-ethylene-responding legumes is *Glycine max* (soybean) in which nodule numbers on ethylene-insensitive mutants and on plants treated with silver ions were similar to those of wild-type plants (Schmidt et al. 1999).

6.2.2 Mutant Analysis and Transgenic Approaches in Plants and Bacteria

The involvement of ethylene in RHC nodulation became clearer by analyzing the symbiotic defects of mutant plants and bacteria. *M. truncatula* plants homozygous for the recessive *sickle* allele had pleiotropic phenotypes, such as delayed petal and leaf senescence and decreased abscission of seedpods and leaves. Seedlings did not show the triple response upon ACC or ethylene treatment, suggesting that the plants were defective in perception of the ethylene signal and that the mutated gene was a component of the ethylene signal transduction pathway. Recently, this gene has been identified as an ortholog of the *Arabidopsis thaliana EIN2* gene (Chan et al. 2005) that encodes a transmembrane protein with an N-terminal domain that shows similarity to the N-ramp family of metal ion transporters (Alonso et al. 1999). Loss-of-function mutations cause complete ethylene insensitivity, showing that EIN2 is a positive regulator in the ethylene signal transduction pathway. In the *sickle* mutant, the number of persistent infections increases considerably in the nodulation zone of the root. Because of a very high number of nodule primordia, this zone becomes swollen and sickle-shaped (Penmetsa and Cook 1997). These observations support the conclusion that ethylene is implicated in arrest of rhizobial infection at the epidermis/cortex interface to control the number of infection events: possibly, a local production of ethylene impedes penetration of ITs into the cortex cells. Abortion of ITs at the epidermis/cortex interface has also been observed in the *S. meliloti*-alfalfa symbiosis (Vasse et al. 1993).

Introduction of a mutated ethylene receptor gene of *Cucumis melo* (melon), *Cm-ERS1/H70A*, into *L. japonicus* conferred reduced ethylene sensitivity, as observed by monitoring root morphology, senescence, and abscission of flowers upon ACC treatment. Control roots were short, thick, and brown, while transgenic roots were white and long. Moreover, petal senescence and detachment were delayed in the transgenic plants when compared to wild-type plants (Nukui et al. 2004). These features probably resulted from a dominant negative mutation that caused deficient ethylene signal transduction (Rodríguez et al. 1999). The phenotype of the transgenic lines was not identical to that of *sickle*, because final nodule number and spatial distribution of the nodules were unaltered. However, both a higher number of ITs and nodule primordia were obtained after bacterial inoculation of transgenic plants. This number of primordia was similar to that of wild-type plants treated with ethylene inhibitors, indicating that the observations are related to the reduced ethylene sensitivity. Independently from the previous work, the *A. thaliana etr1-1* dominant negative ethylene receptor was introduced into *L. japonicus*. Here, the number of nodules increased proportionally to the varying levels of ethylene insensitivity in independent transgenic lines. In lines with higher insensitivity, the increase in nodule number was highest (Guinel and Geil 2002).

Also symbiotic bacteria have been found to modify plant ethylene levels in order to facilitate infection. For instance, *Bradyrhizobium elkanii* produces the phytotoxin rhizobitoxine, an inhibitor of ACC synthase, the enzyme that catalyzes the rate-limiting step of ethylene synthesis. The rate of ethylene synthesis was reduced in siratro plants inoculated with a wild-type, rhizobitoxine-producing strain compared to uninoculated plants, whereas it was equivalent to that of uninoculated plants upon inoculation with a mutant deficient in rhizobitoxine production (Yuhashi et al. 2000). Furthermore, inoculation with the mutant bacteria resulted in a reduced number of nodules compared to that of the wild-type, and the wild-type strain was more competitive for nodulation than the mutant (Yuhashi et al. 2000). Rhizobitoxine-deficient *B. elkanii* strains also formed significantly fewer mature nodules in *Vigna radiata* (mungbean), and ethylene inhibitors partially restored normal nodulation patterns (Duodu et al. 1999).

A different way to interact with ethylene synthesis is the degradation of its direct precursor. For example, certain *Rhizobium leguminosarum* bv. *viciae* strains carry an ACC deaminase gene, the product of which degrades ACC to ammonia and α-ketobutyrate (Ma et al. 2003). These strains are supposed to reduce ethylene biosynthesis in plants by attachment to the root surface and uptake of some of the ACC exuded from the roots. Mutations in the ACC deaminase gene (*acdS*) or the regulatory *lrpL* gene decreased nodulation efficiency of the legume host pea. Moreover, introduction of *acdS* and *lrpL* into *S. meliloti* enhanced nodulation efficiency on alfalfa. These ACC deaminase-producing bacteria were also more competitive than wild-type bacteria because most nodules contained the transgenic bacteria upon co-inoculation with the wild-type strain (Ma et al. 2004).

6.2.3 Ethylene Interferes with NF Signaling Within the Root Hairs

Because ethylene plays a role at the onset of the infection process, its interference with the NF signaling pathway was investigated. Bacteria colonize only developing root hairs in zone I that undergo tip growth and are characterized by a cleared zone in the cytoplasm at the tip. Root hair formation is promoted by ethylene in *A. thaliana* (Tanimoto et al. 1995). Also in vetch, the hormone seems to be a positive regulator of root hair formation, because Ag^+ or AVG completely block root hair growth (Heidstra et al. 1997). On the other hand, root hair deformation, which is induced by bacterial NFs and also involves root hair growth for reinitiation of tip growth, is independent of ethylene in vetch. Treatment with ACC alone failed to induce deformations, whereas addition of AVG or Ag^+ prior to NF application could not inhibit the process. Furthermore, silver treatment of zone I root hairs that normally do not deform, renders them susceptible to NF signaling because the clear zone of the root hair disappeared and their growth was stopped (Heidstra et al. 1997).

NF-induced epidermal responses were further investigated for ethylene interference, such as induction of gene expression and calcium spiking within the root hair, which is characterized by repetitive oscillations in cytosolic calcium (Ehrhardt et al. 1996). Ethylene inhibited NF-dependent *rip1* and *ENOD11* gene expression and interfered with Ca^{++}-spiking (Oldroyd et al. 2001). The percentage of root hair cells that spiked in response to NFs was significantly lower for plants grown on ACC than for those on AVG or *sickle* plants. However, this block of calcium spiking by ACC in wild-type plants was incomplete and could be overcome with a higher concentration of NF. Ethylene application could also block maintenance of the process in wild-type but not in *sickle* plants. Additionally, the frequency of the spikes decreased in the *sickle* mutant when compared to the wild-type situation (Oldroyd et al. 2001). Together these data demonstrate that ethylene regulates a component of the NF signal transduction pathway at or upstream of calcium spiking and defines the sensitivity of the plant to NFs. Clearly, the mechanism involves communication between components of the ethylene-signaling pathway and the NF perception pathway.

6.3 Ethylene is Indispensable for LRB Nodulation

Whereas ethylene plays a negative role in many RHC interactions, it is indispensable for LRB nodulation via crack-entry. This crack-entry invasion, which is an intercellular infection mechanism, has been mainly investigated in the semi-aquatic tropical legume *S. rostrata*. In addition, it has been described during nodulation of *Stylosanthes* sp., *Neptunia* sp., *Mimosa scabrella*, *Aeschynomene afraspera*, and *Chamaecytisus proliferus* (Chandler et al. 1982; de Faria et al. 1988; Alazard and Duhoux 1990; James et al. 1992a, 1992b; Subba-Rao et al. 1995; Vega-Hernández et al. 2001).

6.3.1 Pharmacological Data Show that Ethylene is Needed for Crack-Entry Invasion

When *S. rostrata* roots are grown under hydroponic conditions, i.e., in tubes filled with medium, few root hairs are observed and bulge-like structures occur at the base of lateral roots. Upon addition of purified *A. caulinodans* NFs, the bulges grow out and bushes of distorted axillary root hairs are formed (D'Haeze et al. 2003). Exogenous ethylene, ACC, and ethephon could partially mimic this response and trigger outgrowth of bushes of axillary root hairs that were straight, in contrast to the distorted ones provoked by NFs. On the other hand, AVG or Ag^+ addition prior to NFs blocked the root hair response, indicating that ethylene mediates the axillary root hair outgrowth and might be a NF downstream signal in *S. rostrata* (D'Haeze et al. 2003).

Additionally, ethylene antagonists completely blocked nodulation of hydroponic roots when applied prior to bacterial inoculation. In 50–60% of the roots, LRBs appeared similar to uninoculated roots with neither IP formation nor cell divisions, whereas 40–50% of the LRBs were slightly swollen, probably caused by leaky inhibition. However, IPs could not be observed in these structures. Addition of Ag^+ at different time points before, simultaneously with, or after inoculation, revealed that the inhibitor presumably blocked initiation of nodulation, because addition from day 1 after inoculation already formed a small number of nodules. This number did not increase with time as in control experiments because further initiations were stopped (D'Haeze et al. 2003). On the other hand, although ethylene had no effect on nodulation of hydroponic roots, it induced the formation of lesions and cell death. Based on these observations, which are reminiscent to aerenchyma formation, NF-induced ethylene production might be involved in IP formation, a process that is the primary step for intercellular invasion and depends on NFs (D'Haeze et al. 1998, 2003). Because no nodule primordia could be observed after addition of ethylene antagonists, infection could not be uncoupled from induction of cell division.

6.3.2 Ethylene Mediates the Phenotypic Plasticity in Root Nodule Development

Ethylene requirement in LRB nodulation of *S. rostrata* can be attributed to the plant's adaptation to waterlogging conditions (Goormachtig et al. 2004b). *S. rostrata* requires versatile growth and nodulation capacities to survive and to nodulate in a desert habitat during the rain season. Most species of the genus *Sesbania* develop indeterminate nodules, so does *S. rostrata* on well-aerated roots that are aeroponically grown. Inoculation of roots growing in vermiculite resulted in mature nodules of the indeterminate type with a zonation pattern typical of indeterminate nodules including the persistent apical meristem. Transcripts corresponding to molecular markers for cell division (*Sesro;CycB1;1* and H4-1Sr) were visualized in 30-day-old elongated root nodules in the narrow apical zone of small meristematic cells, clearly proving that the nodules formed under these conditions are of the indeterminate type (Fernández-López et al. 1998). In contrast, hydroponically grown roots carry determinate nodules. The development of these nodules is of a hybrid nature because at the onset of nodule formation different developmental zones are transiently observed, like in indeterminate nodules (Tsien et al. 1983; Duhoux 1984; Ndoye et al. 1994; Goormachtig et al. 1997). However, the meristematic activity is arrested at an early developmental stage (Fernández-López et al. 1998). Ethylene was shown to be a main player in this nodule plasticity. By addition of Ag^+ 5 days after inoculation of hydroponically grown roots, indeterminate nodules were formed, whereas after addition of ACC or ethephon to aerated roots in vermiculite, determinate instead of indeterminate nodules were obtained. These data together provided evidence that the switch in nodule type in *S. rostrata* is mediated by the plant hormone ethylene (Fernández-López et al. 1998). This switch might be due to a negative effect of ethylene on the nodule meristem. Ethylene diffuses a thousand times less well in water than in air and accumulating ethylene might arrest meristematic activity. Under aerated conditions, ethylene might escape more easily, resulting in prolongation of meristematic activity and indeterminate growth. Stem nodules, which develop in the air, have only nodules of the determinate type. The presence of an enhanced gaseous diffusion barrier (James et al. 1998) might lead to an ethylene concentration high enough to block meristem activity.

6.3.3 Ethylene Mediates the Switch from Intercellular to Intracellular Invasion

In addition to the phenotypic plasticity in nodule type, ethylene is involved in determining the infection mechanism in *S. rostrata* (Goormachtig et al. 2004b). In contrast to hydroponic growth during which crack-entry invasion is used, bacteria enter via the RHC process under aeroponic conditions and the corresponding nodules are designated zone I nodules. When AVG or Ag^+ was added to the vermiculite-grown roots 2 days before inoculation, the number of zone I nodules increased, in contrast to a decrease upon ACC addition. Moreover, when vermiculite-grown roots were submerged 24 h or 1 h before inoculation, RHC invasion was completely blocked and no zone I nodules were detected. On the other hand, roots grown hydroponically in the presence of AVG partially had root hairs that could curl and be infected to form zone I nodules (Goormachtig et al. 2004b). Thus, RHC invasion in *S. rostrata* is sensitive to ethylene, and submergence inhibits the invasion of accessible root hairs. Similar findings were made for another water-adapted legume, *Neptunia plena* (Goormachtig et al. 2004b). Plants that are adapted to live under aquatic conditions have developed a mechanism to circumvent ethylene-sensitive RHC nodulation (Goormachtig et al. 2004a).

6.4 Ethylene Determines Nodule Primordium Positioning

Ethylene does not only play a role at the level of infection, the hormone is also implicated in nodule primordium formation and positioning. Analysis of pea mutants confirmed the involvement of ethylene in primordium formation besides its negative role at the epidermis/cortex transition. The phenotype of the pea mutant E2 (*sym5*), which had lost the ability to make nodule primordia but had a normal number of ITs, could be partially rescued by application of silver ions (Fearn and LaRue 1991; Guinel and LaRue 1991). Furthermore, ethylene antagonists restored nodulation in the pea mutant R50 (*sym16*), in which most of the ITs had lost directional growth toward the stele and were arrested in the inner cortex, while nodule primordia were aborted (Guinel and Sloetjes 2000).

Nodule primordia predominantly develop in the root cortex opposite protoxylem poles. This observation led Libbenga et al. (1973) to postulate the presence of transverse gradients of endogenous factors that control induction of cell division in the root, one of which would be ethylene. This hypothesis was supported by experimental data that demonstrated that pea roots grown in the presence of Ag^+ or AVG, had an increased number of primordia developing opposite to protophloem poles, from less than 1% in the non-treated roots up to approximately 10% when the inhibitors were added. Moreover, ACC oxidase transcripts were localized via in situ hybridization in the cell layers opposite protophloem poles (Heidstra et al. 1997). Because the corresponding enzyme converts ACC into ethylene, its location most probably coincides with the actual site of ethylene production. Ethylene produced in the cells opposite phloem poles was proposed to create a gradient that negatively influences primordium formation (Heidstra et al. 1997). Also in the *sickle* mutant of *M. truncatula*, nodule foci surrounded the vascular tissue at all sites. Transverse sections in the nodulation zone showed that nodules were formed at approximately equal frequencies throughout all portions of the root and not only to the protoxylem poles (Penmetsa et al. 2003). Similarly, the number of primordia initiated in-between protoxylem poles increased in transgenic *L. japonicus* containing the *etr1-1* mutant ethylene receptor (Guinel and Geil 2002).

The negative effect on primordium formation might originate from a negative effect on cell division. In *P. sativum*, ethylene has been shown to inhibit cell divisions necessary for growth of the apical hook of etiolated seedlings. Moreover, ethylene retarded cell division in intact root seedlings and prevented lateral bud outgrowth, possibly because of a lack of DNA synthesis (Apelbaum and Burg 1972). Furthermore, when pea plants were treated with exogenous ethylene, not only rhizobial infection was blocked in the outer cortex but also formation of nodule primordia was prevented (Lee and LaRue 1992). Finally, the lack of nodules on thick short roots of *V. sativa* subsp. *nigra* was partially due to decreased nodule primordium formation. Application of the inhibitor AVG restored normal nodulation (Zaat et al. 1989; van Spronsen et al. 1995). These data indicate that, at least in pea, the onset of cell division can be a target for inhibition by ethylene, and that ethylene, produced in the pericycle, plays a role in the spatial control of nodule development by a mechanism that is common to pea, *Medicago*, and *Lotus*.

6.5 Long-Distance Regulation Does Not Involve Ethylene

Nodulation can be controlled locally and ethylene might play a role in this process by restricting nodule primordia and bacterial invasion. The plant also provides a systemic control of nodule numbers via an autoregulatory signal from the shoot that blocks the initiation of new nodule primordia. Because of the link between ethylene, NF signal transduction and nodulation, the hypothesis has been put forward that ethylene would be part of this autoregulatory feedback mechanism (Wood 2001).

The systemic mechanism has been nicely demonstrated by the mutants *har1* in *L. japonicus*, *sym29* in pea, *GmNARK* in soybean, and *sunn* in *M. truncatula*, which are deficient in autoregulation and nodulate all over the root (Krusell et al. 2002; Nishimura et al. 2002; Penmetsa et al. 2003; Searle et al. 2003; Schnabel et al. 2005). Grafting experiments have shown that the plants were mutated in a gene that controlled nodule number from the shoot. This gene codes for a receptor-like kinase with leucine-rich repeats in the extracellular domain (LRR-RLKs) and has a high level of similarity with *CLAVATA1* of *A. thaliana* that negatively regulates formation of shoot and floral meristems by short-distance signaling. In legume nodulation, the gene product is involved in long-distance communication with nodule and lateral root primordia to control the symbiosis (Downie and Parniske 2002). The available literature strongly indicates that ethylene is not involved in the long-distance autoregulatory system. For instance, the *sunn* mutant of *M. truncatula* is normally sensitive to ethylene in comparison to the *sickle* mutant in which ethylene insensitivity is causal to a zone I restricted hyper-nodulation phenotype. Furthermore, both mutated genes have been shown to act in distinct genetic pathways that control nodule number (Penmetsa et al. 2003).

6.6 Conclusions

In many legumes, ethylene has an inhibitory effect on nodulation: nodule numbers decrease by application of exogenous ethylene and increase in the presence of inhibitors of ethylene synthesis or perception. However, the mechanism of ethylene interference with nodulation or the stage at which this occurs may be somewhat different in various plants. Obviously, this variance can correspond to real differences because of specific properties of legumes or their interacting partners, or it can be caused by discrepancies in experimental approaches, such as alternative plant growth conditions, inoculation methods, and pharmacological treatments. An extensive amount of data concerning the location of ethylene action during RHC nodulation has become available. Therefore, several important conclusions can be drawn. Clearly, ethylene acts at the level of bacterial infection; in particular, the epidermis/cortex interface has been described as a target for restriction of bacterial entry, both in *M. truncatula* and pea. Otherwise, it was elegantly reported that epidermal infection of the root hairs is affected by ethylene in *M. truncatula*. The hormone interferes in the NF signal transduction pathway at the level of calcium spiking, one of the earliest responses in the *Rhizobium*-legume symbiosis. Other reports suggest the formation of cytoplasmic bridges in the outer cortex as a possible target for ethylene action, for instance in *V. sativa* and *L. japonicus*. In these species, as well as in pea, ethylene also controls the formation of nodule primordia, which might be due to the general inhibitory effect of ethylene on cell division. This latter feature has also been proposed as the main cause of nodule primordium positioning, which usually takes place opposite protoxylem poles in-between local sites of ethylene production.

Interestingly, when bacteria invade the plant via LRB invasion, ethylene is not inhibitory but absolutely required for nodule initiation. LRB infection has developed as an adaptation to waterlogging and does not involve epidermal responses. Bacteria immediately enter cortical tissue and make IPs by local cell death. Inhibitor studies indicate that the formation of these structures requires the hormone ethylene as a NF downstream signal. Ethylene also determines the switch from RHC to LRB nodulation, and it mediates the phenotypic plasticity in root nodule development in *S. rostrata*. Hence, ethylene is essential for water-tolerant growth and nodulation and it plays a controlling role in nodule positioning and in monitoring bacterial invasion.

Acknowledgements: The authors thank M. De Cock for help in preparing the manuscript. This research was supported by grants from the Interuniversity Poles of Attraction Programme-Belgian Science Policy (P5/13) and by the Research Foundaton-Flanders ("Krediet aan Navorsers" 1.5.088.99.N and 1.5.192.01 N). J.D.H. is a Research Assistant of the Research Foundation-Flanders.

References

Alazard D, Duhoux E (1990) Development of stem nodules in a tropical forage legume, *Aeschynomene afraspera*. J Exp Bot 41:1199–1206

Alonso JM, Hirayama T, Roman G, Nourizadeh S, Ecker JR (1999) EIN2, a bifunctional transducer of ethylene and stress responses in *Arabidopsis*. Science 284:2148–2152

Apelbaum A, Burg SP (1972) Effect of ethylene on cell division and deoxyribonucleic acid synthesis in *Pisum sativum*. Plant Physiol 50:117–124

Brewin NJ (2004) Plant cell wall remodelling in the Rhizobium–legume symbiosis. Crit Rev Plant Sci 23:293–316

Chan PK, Biswas B, Gresshoff P (2005) Screening and characterization for ethylene insensitive mutants. Abstract presented at the 2005 Model Legume Congress, Pacific Grove, CA (USA), 5–9 June 2005, p 85 (#P7)

Chandler MR, Date RA, Roughley RJ (1982) Infection and root-nodule development in *Stylosanthes* species by *Rhizobium*. J Exp Bot 33:47–57

Chen WM, James EK, Prescott AR, Kierans M, Sprent JI (2003) Nodulation of *Mimosa* spp. by the β-proteobacterium *Ralstonia taiwanensis*. Mol Plant-Microbe Interact 16:1051–1061

de Faria SM, Hay GT, Sprent JI (1988) Entry of rhizobia into roots of *Mimosa scabrella* Bentham occurs between epidermal cells. J Gen Microbiol 134:2291–2296

D'Haeze W, Holsters M (2002) Nod factor structures, responses, and perception during initiation of nodule development. Glycobiology 12:79R–105R

D'Haeze W, Gao M, De Rycke R, Van Montagu M, Engler G, Holsters M (1998) Roles for azorhizobial Nod factors and surface polysaccharides in intercellular invasion and nodule penetration, respectively. Mol Plant-Microbe Interact 11:999–1008

D'Haeze W, De Rycke R, Mathis R, Goormachtig S, Pagnotta S, Verplancke C, Capoen W, Holsters M (2003) Reactive oxygen species and ethylene play a positive role in lateral root base nodulation of a semi-aquatic legume. Proc Natl Acad Sci USA 100:11789–11794

Dehio C, de Bruijn FJ (1992) The early nodulin gene *SrEnod2* from *Sesbania rostrata* is inducible by cytokinin. Plant J 2:117–128

Downie JA, Parniske M (2002) Fixation with regulation. Nature 420:369–370

Duhoux E (1984) Ontogénèse des nodules caulinaires du *Sesbania rostrata* (légumineuses). Can J Bot 62:982–994

Duodu S, Bhuvaneswari TV, Stokkermans TJW, Peters NK (1999) A positive role for rhizobitoxine in *Rhizobium*-legume symbiosis. Mol Plant-Microbe Interact 12:1082–1089

Ehrhardt DW, Wais R, Long SR (1996) Calcium spiking in plant root hairs responding to Rhizobium nodulation signals. Cell 85:673–681

Fang Y, Hirsch AM (1998) Studying early nodulin gene *ENOD40* expression and induction by nodulation factor and cytokinin in transgenic alfalfa. Plant Physiol 116:53–68

Fearn JC, LaRue TA (1991) Ethylene inhibitors restore nodulation to *sym* 5 mutants of *Pisum sativum* L. cv Sparkle. Plant Physiol 96:239–244

Ferguson BJ, Mathesius U (2003) Signaling interactions during nodule development. J Plant Growth Regul 22:47–72

Ferguson BJ, Ross JJ, Reid JB (2005) Nodulation phenotypes of gibberellin and brassinosteroid mutants of pea. Plant Physiol 138:2396–2405

Fernández-López M, Goormachtig S, Gao M, D'Haeze W, Van Montagu M, Holsters M (1998) Ethylene-mediated phenotypic plasticity in root nodule development on *Sesbania rostrata*. Proc Natl Acad Sci USA 95:12724–12728

Gage DJ (2004) Infection and invasion of roots by symbiotic, nitrogen-fixing rhizobia during nodulation of temperate legumes. Microbiol Mol Biol Rev 68:280–300

Gage DJ, Margolin W (2000) Hanging by a thread: invasion of legume plants by rhizobia. Curr Opin Microbiol 3:613–617

Geurts R, Bisseling T (2002) *Rhizobium* Nod factor perception and signalling. Plant Cell 14:S239–S249

Goormachtig S, Alves-Ferreira M, Van Montagu M, Engler G, Holsters M (1997) Expression of cell cycle genes during *Sesbania rostrata* stem nodule development. Mol Plant-Microbe Interact 10:316–325

Goormachtig S, Mergaert P, Van Montagu M, Holsters M (1998) The symbiotic interaction between *Azorhizobium caulinodans* and *Sesbania rostrata*. Molecular cross-talk in a beneficial plant-bacterium interaction. In: Biswas BB, Das HK (eds) Plant-microbe interactions. Subcellular biochemistry, vol 29. Plenum Press, New York, pp 117–164

Goormachtig S, Capoen W, Holsters M (2004a) *Rhizobium* infection: lessons from the versatile nodulation behaviour of water-tolerant legumes. Trends Plant Sci 9:518–522

Goormachtig S, Capoen W, James EK, Holsters M (2004b) Switch from intracellular to intercellular invasion during water stress-tolerant legume nodulation. Proc Natl Acad Sci USA 101:6303–6308

Grobbelaar N, Clarke B, Hough MC (1971) The nodulation and nitrogen fixation of isolated roots of *Phaseolus vulgaris* L. III. The effect of carbon dioxide and ethylene. Plant Soil Spec vol 1971:215–223

Guinel FC, Geil RD (2002) A model for the development of the rhizobial and arbuscular mycorrhizal symbioses in legumes and its use to understand the roles of ethylene in the establishment of these two symbioses. Can J Bot 80:695–720

Guinel FC, LaRue TA (1991) Light-microscopy study of nodule initiation in *Pisum sativum* L. cv Sparkle and in its low-nodulating mutant E2 (*sym* 5). Plant Physiol 97:1206–1211

Guinel FC, LaRue TA (1992) Ethylene inhibitors partly restore nodulation of pea mutant E107 (*brz*). Plant Physiol 99:515–518

Guinel FC, Sloetjes LL (2000) Ethylene is involved in the nodulation phenotype of *Pisum sativum* R50 (*sym16*), a pleiotropic mutant that nodulates poorly and has pale green leaves. J Exp Bot 51 885–894

Heidstra R, Yang WC, Yalcin Y, Peck S, Emons A, van Kammen A, Bisseling T (1997) Ethylene provides positional information on cortical cell division but is not involved in Nod factor-induced root hair tip growth in *Rhizobium*-legume interaction. Development 124:1781–1787

Hirsch AM (1992) Developmental biology of legume nodulation. New Phytol 122:211–237

Hirsch AM, Fang Y (1994) Plant hormones and nodulation: what's the connection? Plant Mol Biol 26:5–9

Hirsch AM, Bhuvaneswari TV, Torrey JG, Bisseling T (1989) Early nodulin genes are induced in alfalfa root outgrowths elicited by auxin transport inhibitors. Proc Natl Acad Sci USA 86:1244–1248

Hoffman T, Schmidt JS, Zheng X, Bent AF (1999) Isolation of ethylene-insensitive soybean mutants that are altered in pathogen susceptibility and gene-for-gene disease resistance. Plant Physiol 119:935–949

James EK, Minchin FR, Sprent JI (1992a) The physiology and nitrogen-fixing capability of aquatically and terrestrially grown *Neptunia plena*: the importance of nodule oxygen supply. Ann Bot 69:181–187

James EK, Sprent JI, Sutherland JM, McInroy SG, Minchin FR (1992b) The structure of the nitrogen-fixing root nodules of the aquatic mimosoid legume *Neptunia plena*. Ann Bot 69:173–180

James EK, Minchin FR, Oxborough K, Cookson A, Baker NR, Witty JF, Crawford RMM, Sprent JI (1998) Photosynthetic oxygen evolution within *Sesbania rostrata* stem nodules. Plant J 13:29–38

Kijne JW (1992) The Rhizobium infection process. In: Stacey, G, Burris RH, Evans HJ (eds) Biological nitrogen fixation. Chapman and Hall, New York, pp 349–398

Krusell L, Madsen LH, Sato S, Aubert G, Genua A, Szczyglowski K, Duc G, Kaneko T, Tabata S, de Bruijn F, Pajuelo E, Sandal N, Stougaard J (2002) Shoot control of root development and nodulation is mediated by a receptor-like kinase. Nature 420:422–426

Lee KH, LaRue TA (1992) Exogenous ethylene inhibits nodulation of *Pisum sativum* L. cv Sparkle. Plant Physiol 100:1759–1763

Libbenga KR, Van Iren F, Bogers RJ, Schraag-Lamers MF (1973) The role of hormones and gradients in the initiation of cortex proliferation and nodule formation in *Pisum sativum* L. Planta 114:29–39

Lievens S, Goormachtig S, Den Herder J, Capoen W, Mathis R, Hedden P, Holsters M (2005) Gibberellins are involved in nodulation of *Sesbania rostrata* nodulation. Plant Physiol 139:1366–1379

Ma W, Guinel FC, Glick BR (2003) *Rhizobium leguminosarum* biovar viciae 1-aminocyclopropane-1-carboxylate deaminase promotes nodulation of pea plants. Appl Environ Microbiol 69:4396–4402

Ma W, Charles TC, Glick BR (2004) Expression of an exogenous 1-aminocyclopropane-1-carboxylate deaminase gene in *Sinorhizobium meliloti* increases its ability to nodulate alfalfa. Appl Environ Microbiol 70:5891–5897

Mathesius U, Schlaman HRM, Spaink HP, Sautter C, Rolfe BG, Djordjevic MA (1998) Auxin transport inhibition precedes root nodule formation in white clover roots and is regulated by flavonoids and derivatives of chitin oligosaccharides. Plant J 14:23–34

Mathesius U, Charon C, Rolfe BG, Kondorosi A, Crespi M (2000) Temporal and spatial order of events during the induction of cortical cell divisions in white clover by *Rhizobium leguminosarum* bv. *trifolii* inoculation or localized cytokinin addition. Mol Plant-Microbe Interact 13:617–628

Moulin L, Munive A, Dreyfus B, Boivin-Masson C (2001) Nodulation of legumes by members of the β-subclass of Proteobacteria. Nature 411:948–950

Ndoye I, de Billy F, Vasse J, Dreyfus B, Truchet G (1994) Root nodulation of *Sesbania rostrata*. J Bacteriol 176:1060–1068

Nishimura R, Hayashi M, Wu G-J, Kouchi H, Imaizumi-Anraku H, Murakami Y, Kawasaki S, Akao S, Ohmori M, Nagasawa M, Harada K, Kawaguchi M (2002) HAR1 mediates systemic regulation of symbiotic organ development. Nature 420:426–429

Nukui N, Ezura H, Yuhashi KI, Yasuta T, Minamisawa K (2000) Effects of ethylene precursor and inhibitors for ethylene biosynthesis and perception on nodulation in *Lotus japonicus* and *Macroptilium atropurpureum*. Plant Cell Physiol 41:893–897

Nukui N, Ezura H, Minamisawa K (2004) Transgenic *Lotus japonicus* with an ethylene receptor gene *Cm-ERS1/H70A* enhances formation of infection threads and nodule primordia. Plant Cell Physiol 45:427–435

Oldroyd GED, Engstrom EM, Long SR (2001) Ethylene inhibits the Nod factor signal transduction pathway of *Medicago truncatula*. Plant Cell 13:1835–1849

Pawlowski K, Bisseling T (1996) Rhizobial and actinorhizal symbioses: what are the shared features? Plant Cell 8:1899–1913

Penmetsa RV, Cook DR (1997) A legume ethylene-insensitive mutant hyperinfected by its rhizobial symbiont. Science 275:527–530

Penmetsa RV, Frugoli JA, Smith LS, Long SR, Cook DR (2003) Dual genetic pathways controlling nodule number in *Medicago truncatula*. Plant Physiol 131:998–1008

Peters NK, Crist-Estes DK (1989) Nodule formation is stimulated by the ethylene inhibitor aminoethoxyvinylglycine. Plant Physiol 91:690–693

Rodríguez FI, Esch JJ, Hall AE, Binder BM, Schaller GE, Bleecker AB (1999) A copper cofactor for the ethylene receptor ERT1 from *Arabidopsis*. Science 283:996–998

Schmidt JS, Harper JE, Hoffman TK, Bent AF (1999) Regulation of soybean nodulation independent of ethylene signaling. Plant Physiol 119:951–959

Schnabel E, Journet E-P, de Carvalho-Niebel F, Duc G, Frugoli J (2005) The *Medicago truncatula SUNN* gene encodes a *CLV1*-like leucine-rich repeat receptor kinase that regulates nodule number and root length. Plant Mol Biol 58:809–822

Searle IR, Men AE, Laniya TS, Buzas DM, Iturbe-Ormaetxe I, Carroll BJ, Gresshoff PM (2003) Long-distance signaling in nodulation directed by a CLAVATA1-like receptor kinase. Science 299:109–112

Sprent JI (2002) Nodulation in legumes. Royal Botanical Gardens, Kew

Subba-Rao NS, Mateos PF, Baker D, Pankratz HS, Palma J, Dazzo FB, Sprent JI (1995) The unique root-nodule symbiosis between *Rhizobium* and the aquatic legume, *Neptunia natans* (L. f.) Druce. Planta 196:311–320

Sy A, Giraud E, Jourand P, Garcia N, Willems A, de Lajudie P, Prin Y, Neyra M, Gillis M, Boivin-Masson C, Dreyfus B (2001) Methylotrophic *Methylobacterium* bacteria nodulate and fix nitrogen in symbiosis with legumes. J Bacteriol 183:214–220

Syōno K, Torrey JG (1976) Identification of cytokinins of root nodules of the garden pea, *Pisum sativum* L. Plant Physiol 75:602–606

Tanimoto M, Roberts K, Dolan L (1995) Ethylene is a positive regulator of root hair development in *Arabidopsis thaliana*. Plant J 8:943–948

Timmers ACJ, Auriac M-C, Truchet G (1999) Refined analysis of early symbiotic steps of the *Rhizobium-Medicago* interaction in relationship with microtubular cytoskeleton rearrangements. Development 126:3617–3628

Tsien HC, Dreyfus BL, Schmidt EL (1983) Initial stages in the morphogenesis of nitrogen-fixing stem nodules of *Sesbania rostrata*. J Bacteriol 156:888–897

van Spronsen PC, van Brussel AAN, Kijne JW (1995) Nod factors produced by *Rhizobium leguminosarum* biovar *viciae* induce ethylene-related changes in root cortical cells of *Vicia sativa* spp. *nigra*. Eur J Cell Biol 68:463–469

van Spronsen PC, Grønlund M, Pacios Bras C, Spaink HP, Kijne JW (2001) Cell biological changes of outer cortical root cells in early determinate nodulation. Mol Plant-Microbe Interact 14:839–847

Vasse J, de Billy F, Camut S, Truchet G (1990) Correlation between ultrastructural differentiation of bacteroids and nitrogen fixation in alfalfa nodules. J Bacteriol 172:4295–4306

Vasse J, de Billy F, Truchet G (1993) Abortion of infection during the *Rhizobium meliloti*—alfalfa symbiotic interaction is accompanied by a hypersensitive reaction. Plant J 4:555–566

Vega-Hernández MC, Pérez-Galdona R, Dazzo FB, Jarabo-Lorenzo A, Alfayate MC, León-Barrios M (2001) Novel infection process in the indeterminate root nodule symbiosis between *Chamaecytisus proliferus* (tagasaste) and *Bradyrhizobium* sp. New Phytol 150:707–721

Wood NT (2001) Nodulation by numbers: the role of ethylene in symbiotic nitrogen fixation. Trends Plant Sci 6:501–502

Yuhashi K-I, Ichikawa N, Ezura H, Akao S, Minakawa Y, Nukui N, Yasuta T, Minamisawa K (2000) Rhizobitoxine production by *Bradyrhizobium elkanii* enhances nodulation and competitiveness of *Macroptilium atropurpureum*. Appl Environ Microbiol 66:2658–2663

Zaat SAJ, Van Brussel AAN, Tak T, Lugtenberg BJJ, Kijne JW (1989) The ethylene-inhibitor aminoethoxyvinylglycine restores normal nodulation by *Rhizobium leguminosarum* biovar. *viciae* on *Vicia sativa* subsp. *nigra* by suppressing the 'Thick and short roots' phenotype. Planta 177:141–150

7 The Role of Ethylene in the Regulation of Stem Gravitropic Curvature

Marcia A. Harrison

7.1 The Modulating Role of Ethylene on Stem Gravitropic Curvature

Ethylene's involvement in a plant's ability to orient itself to a gravitational field has been recognized for almost a century. In 1910, Neljubov identified ethylene as the active component of illuminating gas (used to heat greenhouses) that affected plant growth and caused senescence. He noted that pea epicotyls exposed to high levels of ethylene (1) became oriented horizontally, (2) showed reduced length and (3) increased in diameter. Hence, ethylene's effect on the plants' physiology became known as the "triple response", which has become extensively used as an ethylene bioassay as well as a diagnostic phenotype for screening ethylene overproducing or insensitive mutants. Other early investigations indirectly demonstrated ethylene's involvement in plant responses to gravity (reviewed by Abeles 1973). For example, horizontally placed pineapple plants flower earlier than upright plants. Since ethylene induces flowering in bromeliads, early flowering was attributed to being a side effect of the increase in ethylene production caused by horizontal reorientation of the plants. In another experiment, horizontally oriented plants that were continually rotated on a clinostat exhibited physiological responses such as downward leaf bending and growth inhibition, reactions known to be caused by increased ethylene production.

Decades of research on ethylene's involvement in the gravitropic response of plant stems have produced contradictory data concerning its effect on curvature. Plant tissues displaying different sensitivity to ethylene, and growth conditions that are likely to alter ethylene production often confound meaningful data comparison. The characterization of ethylene mutants along with genetic and molecular approaches using *Arabidopsis* allows for new insights into the role of ethylene in gravitropism. This chapter will present studies that characterize the regulatory role of ethylene during gravitropic curvature in different types of plant stems and inflorescence stalks, and will also discuss the current understanding of ethylene cross-talk with auxin relative to the regulation of stem gravitropism.

Department of Biological Sciences, Marshall University, Huntington, WV 25755

Ethylene Action in Plants
(ed. by N.A. Khan)
© Springer-Verlag Berlin Heidelberg 2006

7.2 Overview of Curvature Kinetics for Light-Grown Compared to Dark-Grown Tissues

Studies of gravitropic curvature reveal different kinetics for the response of light-grown stem tissues than etiolated tissues. For example, in light-grown tomato plants, stems exhibit strong upward curvature 30 min after horizontal reorientation that continues until an angle of greater than 90 ° (an overshoot) is reached. At this point, the stems are once again stimulated by gravity, and the overshoot is corrected (Fig. 7.1) (Harrison and Pickard 1986). Kinetic studies of gravitropic curvature in tomato hypocotyls demonstrate the variability of the response, where individual hypocotyls differ in the degree of overshoot (Fig. 7.1A). When the curvature measurements are averaged, upward curvature appears to proceed in a constant linear fashion past 90 °, reaching an overshoot by 3 h stimulation (Fig. 7.1B). This type of response has also been reported for cocklebur stems (Wheeler et al. 1986), sunflower hypocotyls and etiolated maize coleoptiles (Firn and Digby 1979), and inflorescence stalks of *Arabidopsis* (Fukaki et al. 1996; Masson et al. 2002) (Fig. 7.2). In *Arabidopsis* inflorescence stalks, curvature begins in the subapical region, where the highest elongation rate occurs (Fukaki et al. 1996). Curvature continues until the tip overshoots to as much as 150 °, then reverses, causing the initial curving zone to straighten, while curvature continues in a more basal stem section. This straightening of curved tissue and a shift in the locus of curvature is termed counter-reaction or autotropic straightening (Firn and Digby 1979; Pickard 1985; Stankovic et al. 1998). By 24 h of stimulation, the locus of curvature is closer to the base of the stalk than to its apex which is fully upright (Fukaki et al. 1996). These results indicate that the entire elongation zone (the distal 4 cm) of the *Arabidopsis* inflorescence stalk is involved in the response, rather than only a distinct motor region (such as the distal elongation zone described for roots).

In most etiolated stems, linear upward curvature begins in the sub-apical region, but slows prior to attaining vertical orientation. As this slowing occurs, the locus of curvature shifts toward the base of the stem while the tip simultaneously straightens, displaying early onset of the counter-reaction. In gravitropic curvature of etiolated pea epicotyls there is little upward curvature until 15 min after horizontal reorientation, after which upward curvature proceeds at a rate of 0.55 ° per min (Fig. 7.3). After 2 h of stimulation, curvature slows to a rate of 0.06 ° per minute, and the plants do not reach fully upright orientation even after 7 h. The extent of counter-reaction varies from species to species, but a net slowing of the rate of curvature prior to 90 ° orientation is observed in many other etiolated stems, including oat coleoptiles (Pickard 1973), maize coleoptiles (Bandurski et al. 1984), cucumber hypocotyls (MacDonald et al. 1983; Cosgrove 1990), tomato hypocotyls (Madlung et al. 1999), cress hypocotyls (Hart and MacDonald 1981; MacDonald et al. 1983), and *Arabidopsis* hypocotyls (Kiss et al. 1999).

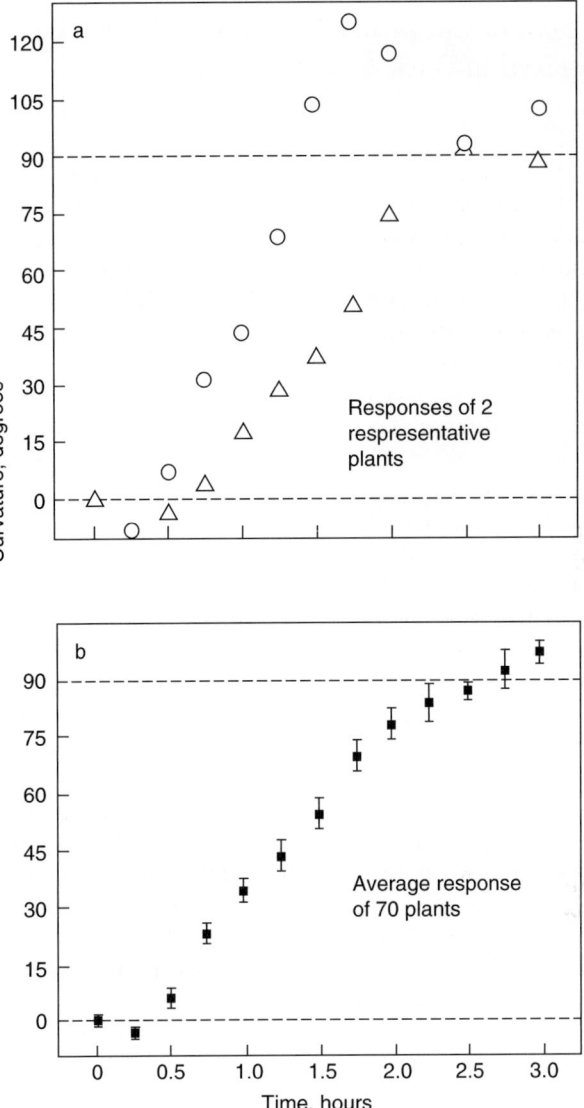

Fig. 7.1. Representative individual time courses of gravitropism **a** and the average time course for 70 hypocotyls **b** of 7-day-old light-grown tomato seedlings (from Harrison and Pickard 1986; copyrighted by the American Society of Plant Biologists and is reprinted with permission)

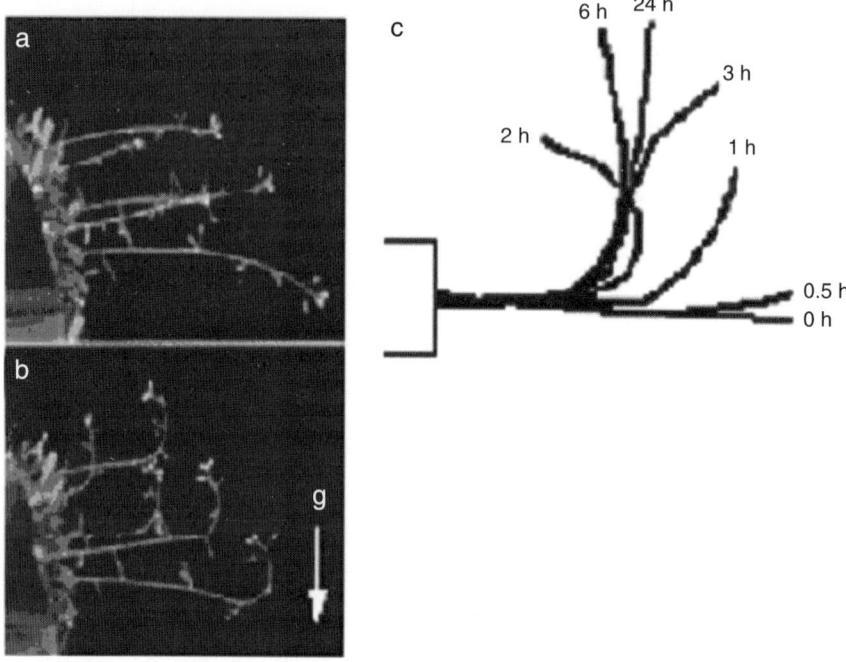

Fig. 7.2. Shoot gravitropism of inflorescence stalks of *Arabidopsis* showing plants at 0 **a** and 90 min **b** after horizontal reorientation. Decapitated inflorescence stalk segments of *Arabidopsis* overshoot 90° by 2-h post-stimulation **c**. For this experiment, excised stem segments were placed horizontally in a gel block used as a holder (shown as a *white rectangle*) at time 0 h. The shape of the stem was traced at the indicated times. The basal side of the stem segment was embedded in the gel block; the apical side was free to move. The *arrow* indicates the direction of gravity (g) (from Masson et al. 2002; copyrighted by the American Society of Plant Biologists and is reprinted with permission)

Therefore, differences in the kinetics of autotropic straightening may be dependent upon the plant's lighting regime.

7.3 Ethylene Production Increases after Horizontal Reorientation and during Curvature

The differential growth that causes the upward curvature of stems in response to gravity is regulated by an auxin gradient in the stem (Masson et al. 2002). However, there is substantial evidence of increased ethylene production during gravitropic curvature (Wright et al. 1978; Clifford et al. 1983; Wheeler et al. 1986; Philosoph-Hadas et al. 1996; Friedman et al. 1998; Madlung et al. 1999; Steed et al. 2004). In dark-grown pea seedlings, ethylene production increases during gravitropic curvature by 65 min after horizontal placement, and

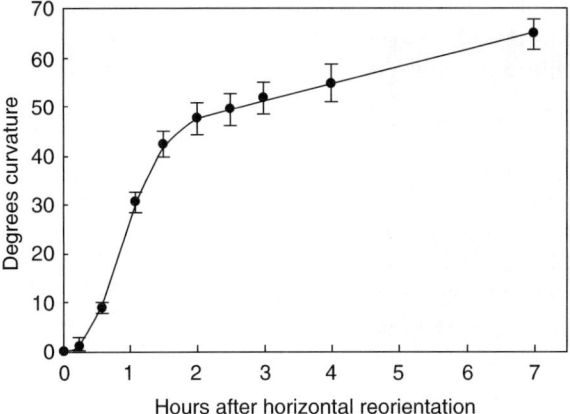

Fig. 7.3. Time course of gravitropic curvature in 5-day-old etiolated pea epicotyls. Means ± SE. $n=16$ plants (B. Pickard and M. Harrison, unpublished)

remains higher than the zero time point at the 120-min time point (Fig. 7.4). Thus, the increase in ethylene production occurs after the onset of upward curvature. Ethylene production is still greater than the zero time point at the beginning of the counter-reactive phase, which begins by 2 h after horizontal reorientation (see Fig. 7.3). Ethylene production is often reported to be greater

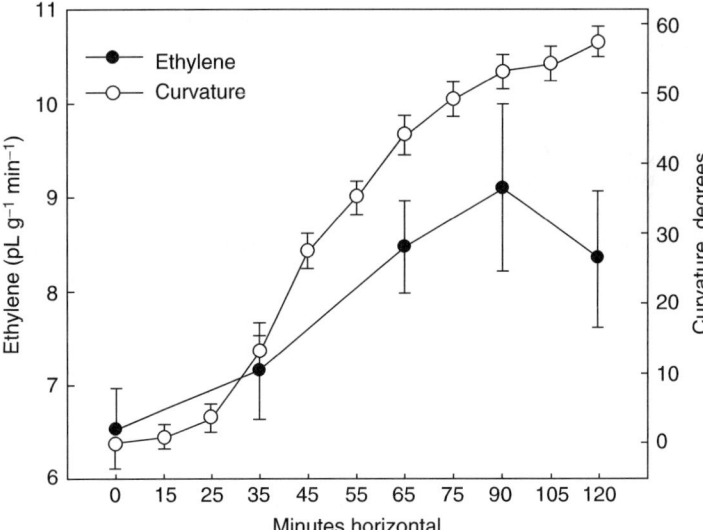

Fig. 7.4. Ethylene production and gravitropic curvature rates in etiolated pea epicotyls after horizontal reorientation. In vivo ethylene measurements from subapical epicotyl sections enclosed in vials were determined by gas chromatography. Means ± SE. $n = 8$ vials for ethylene measurements, and eight sets of 20 plants for curvature analysis (adapted from Figure 1 in Steed et al. 2004 with kind permission of Springer Science and Business Media)

in the lower flank of horizontally placed tissue compared to the upper flank (Abeles and Rubinstein 1964; Wright et al. 1978; Clifford et al. 1983; Wheeler et al. 1986; Philosoph-Hadas et al. 1996). Using dandelion peduncle, Clifford et al. (1983) found a threefold increase in ethylene after horizontal orientation, accompanied by a significant increase in ethylene in the lower flanks during the 4–6 h interval where a curvature of 42.7 ° is obtained. Four hours after horizontal reorientation, the lower halves of snapdragon floral spikes produce over three times more ethylene than the upper halves throughout the length of the spike (Philosoph-Hadas et al. 1996). The greatest ethylene production is found in the region 5–10 cm below the apex, coinciding with the location of the bending zone. The level of the ethylene precursor 1-aminocyclopropane-1-carboxylic acid (ACC) also significantly increases in the lower flanks; supporting the conclusion that ethylene biosynthesis is stimulated in the lower flank of these stalks during curvature.

7.4 Differing Sensitivities to Exogenous Ethylene May Alter the Gravitropic Response

When the effect of exogenous ethylene application is evaluated for different tissues, a wide variation in the sensitivity to ethylene is observed. Etiolated stems such as pea epicotyls are very sensitive to low ethylene levels (20–100 nL L^{-1}). In these tissues, low exogenous levels appreciably inhibit curvature. In etiolated *Arabidopsis* hypocotyls, as little as 0.2 nL L^{-1} exogenous ethylene can cause rapid transient growth inhibition (Binder et al. 2004). In contrast, light-grown tissues exhibit greatly reduced sensitivity to ethylene levels. In light-grown tomato hypocotyls, only high exogenous ethylene levels of 50 and 100 µL L^{-1} inhibit curvature (27% inhibition) (Harrison and Pickard 1986). Wheeler et al. (1986) also observed inhibition of curvature at high ethylene levels of 10 and 100 µL L^{-1} for cocklebur vegetative stems. A more recent study shows stimulation of curvature by application of exogenous ethylene when *Arabidopsis* inflorescence stalks or hypocotyls are pre-exposed to ethylene prior to horizontal placement (Lu et al. 2002). Therefore, differences in the reported effects of ethylene on stem gravitropism vary depending on the relative sensitivity of the tissue and the exposure method.

7.4.1 Evidence of a Stimulatory Role for Ethylene in Stem Gravitropism

In light-grown tissue, a positive regulatory role for ethylene in stem gravitropism has been documented by several studies (Wheeler et al. 1986; Golan et al. 1996; Philosoph-Hadas et al. 1996; Lu et al. 2002). While ethylene acts as an inhibitor of cellular growth in many tissues, ethylene is also known to stimulate growth in specific tissues such as aquatic stems and grass tiller internodes (Abeles et al. 1992), and under certain growth conditions as demonstrated for *Arabidopsis* seedlings (Smalle et al. 1997). The application of 10 µL L^{-1} ethylene to light-grown *Arabidopsis* seedlings induces hypocotyl elongation, especially when the seedlings are grown in a nutrient-poor, sucrose-free medium (Smalle et al. 1997). Thus, depending upon the tissue type and the experimental conditions, the increase in ethylene production that occurs after the horizontal reorientation of stems might contribute to cellular elongation, and not necessarily act as an inhibitor.

Much of the early evidence used to support a positive role for ethylene in gravitropism comes from studies employing exogenous applications of substances that inhibit either ethylene biosynthesis or a plant's response to ethylene. In etiolated tomato hypocotyls, application of the competitive ethylene antagonist 2,5-norbornadiene (NBD) to wild-type plants interferes with the tissue's response to gravity, inhibiting curvature (by 53%) but not elongation at 20 h after gravistimulation (Madlung et al. 1999). Other studies report that ethylene inhibitors reduced curvature in dandelion peduncles (Clifford et al. 1983), vegetative stems of 30 to 40-day-old cocklebur and tomato plants (Wheeler et al. 1986), and snapdragon floral spikes (Philosoph-Hadas et al. 1996). Agents that disrupt the formation of the ethylene gradient across snapdragon floral spikes also greatly reduce curvature in this tissue (Philosoph-Hadas et al. 1996). Thus, for snapdragon spikes, the ethylene gradient is required to produce the differential growth that causes upward curvature. However, inhibitor studies are problematic and inconsistent since these chemicals may cause non-specific effects, especially when used at higher concentrations (e.g., AVG inhibits protein synthesis at levels of 1 mM or higher, Mattoo et al. 1979).

Indirect evidence of ethylene enhancement of the gravitropic response is also observed in *Arabidopsis*, where cytokinin-induced ethylene production restores a normal gravitropic response in red light-grown seedlings that otherwise display random orientation relative to gravity (Golan et al. 1996). Curvature in light-grown hypocotyls and inflorescence stalks is significantly enhanced when the tissue is given both long-term pre-exposure to ethylene and continuous exposure during curvature (Lu et al. 2002). In these experiments, differing ethylene exposure times mimic stress conditions that induce transient ethylene production. In intact inflorescence stalks, a 12-h pretreatment and continued exposure to 0.1–10 µL L^{-1} ethylene stimulates curvature to 120 ° by 2 h, compared to 60 ° in air controls. Interestingly, this ethylene response is similar in pattern to that noted by Fukaki et al. (1996) where decapitated, leafless *Arabidopsis* inflorescence stalks show increased curvature and greater tip overshoot compared to intact controls. It is likely that the process of decapitation and leaf excision stimulates wound ethylene production, thus altering curvature kinetics. In light-grown hypocotyls, only long-term pretreatment (48 h with 5 µL L^{-1}) and continuous ethylene exposure stimulate curvature to 50 ° by 5 h, whereas little curvature response is observed in air controls or with a short-term pre-exposure.

Mutations that cause an in vivo increase in ethylene production support results obtained using exogenous ethylene application. Inflorescence stalks of the *Arabidopsis eto1-1* (ethylene overproducer) mutant exhibit increased curvature similar to that obtained by ethylene pretreatment (Lu et al. 2002). Development of T-DNA insertion mutants of *Arabidopsis* has made it possible to investigate the role of individual ACC synthase (ACS) isoforms (responsible for ACC production) in regulating the gravitropic response. Seedlings with a T-DNA insertion into the *ACS4* gene or with the *eto2* mutation (on *ACS5*) exhibit a higher degree of curvature than wild-type plants by 7 h horizontal placement (Fig. 7.5A). Both the *acs4* and *eto2* mutants also exhibit an increase in ethylene production relative to wild-type plants (Fig. 7.5B). The dominant mutation that results in *eto2* causes a truncated C-terminal domain in ACS5. It is proposed that the C-terminal domain has a site targeted for proteolytic activity that is absent in *eto2*. This would result in the increased stability of the enzyme by reducing its degradation, thus

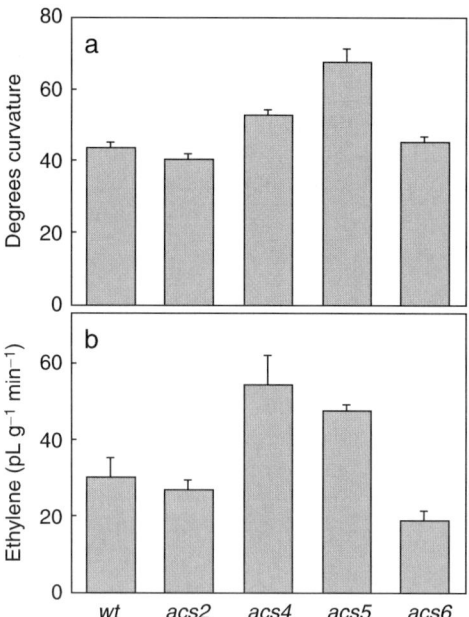

Fig. 7.5. Gravitropic curvature and ethylene production in etiolated hypocotyls of 3-day-old dark-grown *Arabidopsis* wild type compared to *eto2* mutation of *ACS5* (CS8059), or T-DNA insertion mutants *acs2*, *acs4*, and *acs6* (SALK_025672, SALK_054951, and SALK_054467, respectively) 7 h after horizontal reorientation. Seeds for wild-type and SALK Institute T-DNA insertion lines for the *ACS* genes (produced by Joseph R. Ecker and the Salk Institute Genomic Analysis Laboratory) were obtained from the *Arabidopsis* Biological Resource Center, The Ohio State University. Means ± SE. $n=4$ vials for ethylene measurements, and $n= 35-70$ plants for curvature analysis (M. Harrison, unpublished).

leading to ongoing synthesis of ACC (Chae et al. 2003). The T-DNA insertion that results in the *asc4* mutant is also located in the C-terminal exon and may produce a truncated enzyme with increased stability. T-DNA insertion mutants *acs2* and *acs6* did not exhibit a change in curvature, and ethylene production was reduced in the *acs6* mutant compared to wild-type plants. The insertions that create these mutants occur within their first exons potentially preventing expression of the enzyme. Although these studies used etiolated hypocotyls that are reported to be very sensitive to low levels of ethylene (Binder et al. 2004), the results indicate a stimulatory role for ethylene in hypocotyl gravitropic response. It is suggested that both the dim green light used for image capture and the nutrient conditions may be the cause of the stimulatory results. This is supported by studies showing that short-term exposure to green light stimulates hypocotyl elongation (Folta 2004), and that seedlings grown under low nutrient conditions are stimulated by exogenous ethylene (Smalle et al. 1997). Thus, this set of experimental variables may produce a growth condition in which ethylene stimulates growth in hypocotyls and promotes upward curvature.

Mutants insensitive to ethylene often exhibit slow curvature kinetics, supporting a stimulatory role for ethylene in the response. For example, the *Arabidopsis* ethylene insensitive mutant *etr1* (ethylene response) exhibits little ethylene-induced hypocotyl curvature and shows delayed curvature in inflorescence stalks (Lu et al. 2002). The tomato *nr* (never ripe) ethylene insensitive mutant exhibits an increased lag time of 45 min prior to upward curvature, after which the curvature reaches wild type response. These results indicate that tissue sensitivity to ethylene may alter the timing of upward curvature.

7.4.2 Ethylene's Role in Slowing Gravitropic Curvature

Although there is increasing evidence of a stimulatory role for ethylene in regulating gravitropism, there is also strong evidence of ethylene inhibition of the response. In light grown plants, higher levels of ethylene are required to slow gravitropic curvature. However, in etiolated stems, which are especially sensitive to exogenous ethylene, low concentrations are sufficient to reduce curvature. For example, in etiolated pea epicotyls, low exogenous ethylene levels (20 and 50 nL L^{-1}) cause a 50% decrease in curvature at 7 h post-stimulation, while 100 nL L^{-1} exogenous ethylene resulted in a lack of upward curvature, causing the stems to be essentially horizontal (Fig. 7.6). Exogenous application of the competitive ethylene inhibitor NBD causes an increase in net curvature and prevents the severe curvature inhibition otherwise caused by treatment with 100 nL L^{-1} ethylene (Fig. 7.6). Madlung et al. (1999) noted inhibition of curvature in etiolated tomato hypocotyls in the presence of 0.01–1.0 μL L^{-1} ethylene, which represents levels that do not inhibit stem elongation. The tomato ethylene mutant *epi* (epinastic, ethylene

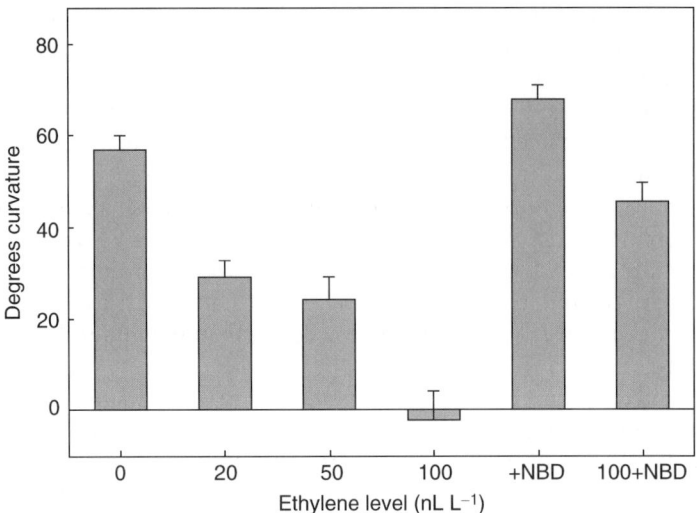

Fig. 7.6. Effect of exogenous ethylene and an ethylene inhibitor (NBD) on gravitropic curvature in etiolated pea epicotyls. NBD was supplied at a concentration of 100 µL L^{-1}. Curvature measurements were taken 7 h after horizontal reorientation. Means ± SE. n=26–35 seedlings (M. Harrison and B. Pickard, unpublished)

overproducing) exhibits altered gravitropic response kinetics compared to wild-type hypocotyls. While wild-type hypocotyls reach 50 ° curvature by 60 min, then slow with a fairly strong counter-reaction, *epi* hypocotyls curve slowly for the first 12 h of stimulation, but reaches wild-type curvature by 24 h (Madlung et al. 1999). Thus, in etiolated stems, conditions or mutations that result in increased ethylene production may act to slow or prevent upward curvature.

7.5 Ethylene and Auxin Cross-Talk in the Regulation of Gravitropic Curvature

Straightening of the subapical portion of the stem is a normal occurrence during gravitropism and requires coordinated changes in cellular growth. Firn and Digby (1979) note that the timing of autotropic straightening contributes to the slowing of curvature prior to reaching 90 °, and to the correction of overshoot. These changes may be caused by differential levels of, or sensitivity to, auxin and ethylene along the stem during curvature. Pickard (1985) proposed that ethylene and auxin "work as balancing members of a feedback system in which IAA (indole-3-acetic acid, an auxin) stimulates the synthesis of ethylene, and ethylene inhibits the synthesis and often the transport of IAA". The cross-talk between the two hormones is likely to be a key regulator of both gravitropic response kinetics and the counter-reaction.

In vertical stems, auxin transport is basipetal. Upon horizontal reorientation of plants, transport shifts laterally, resulting in auxin accumulation in the lower side of the stem. Experiments that track the movement of radiolabeled IAA show that it migrates into the lower flanks of horizontally placed stems minutes after horizontal reorientation (Bandurski et al. 1984; Harrison and Pickard 1989). In tomato hypocotyls, an asymmetric IAA accumulation is measured within 10 min of reorientation (Harrison and Pickard 1989). By 25 min after reorientation, during rapid curvature of the hypocotyls, there is 2.5 times the amount of IAA in the lower flanks than the upper flanks of these hypocotyls.

Auxin transport is controlled by a system of protein influx and efflux carriers. One auxin efflux carrier, PIN3 (pin-formed), is primarily located in the lateral cell surfaces of gravity-sensing tissues. There is evidence from studies of *Arabidopsis* roots that PIN3 relocalizes to the lower membrane surface when the plant is reoriented horizontally (Friml et al. 2002). Plants with mutations of auxin efflux carriers exhibit varying degrees of altered gravitropism in both hypocotyls and roots (Masson et al. 2002). Therefore, localization of these efflux carriers is proposed to be a component of the regulatory system that transports auxin to the lower flank of gravistimulated organs.

The increase in ethylene that occurs in many stems after horizontal reorientation is probably a result of several factors. During the gravitropic response, horizontal placement of the plant tissue causes mechanical perturbation that triggers stress-induced ethylene production. Fine kinetic analysis of ethylene production shows a short transient ethylene burst within minutes after horizontal reorientation of tomato seedlings (Harrison and Pickard 1984). Auxin is also a major contributor to the increase in ethylene levels. Since higher levels of auxin are known to stimulate ethylene biosynthesis, it has often been suggested that auxin accumulation during gravitropism induces ethylene biosynthesis and is the cause of the ethylene gradient between the upper and lower halves of curving stems (Abeles et al. 1992).

In an auxin-ethylene cross-talk model proposed by Swarup et al. (2002), ethylene positively affects auxin accumulation and transport (Fig. 7.7). In this model, ethylene signal transduction leads to the up-regulation of HLS (HOOKLESS, an acetyltransferase), which is involved in regulating differential growth associated with maintenance of the apical hook in etiolated *Arabidopsis* seedlings. HLS affects auxin concentration in the cell by regulating auxin accumulation and/or transport. Although apical hook development is a distinct developmental process from gravitropism, there may be common mechanisms involved in the regulation of differential growth, especially for dark-grown stem tissues in which increased ethylene promotes strong hook formation and reduced stem growth as seedlings push through soil (Raz and Ecker 1999; de Grauwe et al. 2005).

Applying the auxin-ethylene cross-talk model to gravitropism, short-term mechanical stress-induced ethylene production may contribute to the accumulation of auxin in the lower flanks of horizontally reoriented stems, by either increased accumulation or transport (Fig. 7.7). This accumulated auxin drives ethylene biosynthesis increasing ethylene levels after the onset of upward curvature. In etiolated hypocotyls, cellular elongation is inhibited by very low levels of ethylene. Therefore, the resulting increase in ethylene may serve to inhibit elongation in the lower side of the hypocotyl and thus slow upward curvature prior to the tip reaching 90°. According to the model, ethylene in the lower flank of a horizontally reoriented stem may bind to the ETR receptor and continue to drive auxin accumulation and ethylene production in the lower side of the stem, thus further inhibiting growth to a point where the upper side has more rapid growth and the stem begins to straighten. High levels of ethylene may also reduce tissue sensitivity to ethylene, reducing growth inhibition over time. Tissue sensitivity can be controlled by the ethylene receptor system. In *Arabidopsis*, five ethylene receptors are found at varying concentrations in different tissues, allowing tissues to exhibit differing sensitivity to ethylene. Three of the receptors in *Arabidopsis* are induced

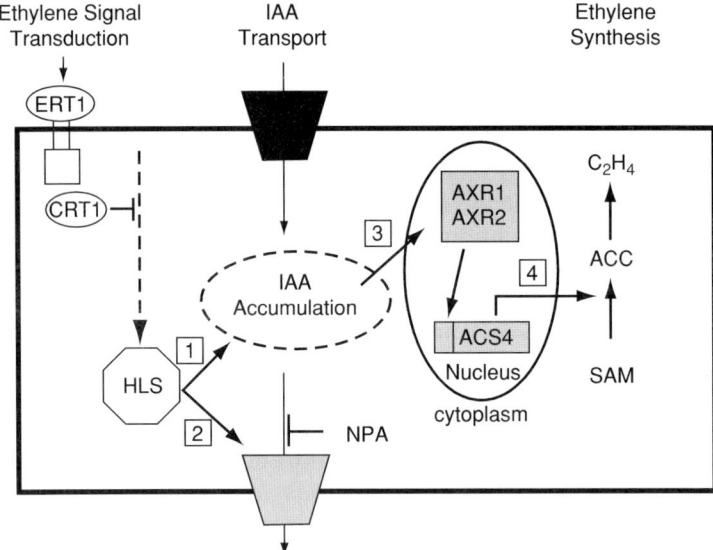

Fig. 7.7. Auxin-ethylene cross-talk in plant cells. In this model, ethylene binds to a membrane-bound ethylene receptor (ETR1, ethylene response), inducing signal transduction via the relay protein CTR1 (constitutive triple response), causing increased expression of HLS (hookless). HLS regulates (1) IAA accumulation and transport (Lehman et al. 1996) and/or (2) the activity of an auxin transporter that is sensitive to NPA (1-naphthylphthalamic acid, an auxin transport inhibitor) (Kieber 1997). Accumulated auxin up-regulates (3) AXR1-(AUXIN RESISTANT) and AXR2-dependent expression of (4) ACC synthase 4 (*ACS4*) (Abel et al. 1995) (from Figure 3 in Swarp et al. 2002 with kind permission of Springer Science and Business Media)

by ethylene (Chen et al. 2005). Since the receptors negatively regulate ethylene signal transduction, an increased number of receptors within a cell may account for reduced tissue sensitivity. Thus, when ethylene production in a tissue increases, tissue sensitivity is reduced and lateral auxin transport and accumulation slows. This may also allow more basally directed movement of auxin relative to lateral movement, causing a migration of curvature to a more basal section of the stem, as is usually observed in gravistimulated stems.

Light-grown tissues are less sensitive to ethylene, so accumulated auxin and increased ethylene production does not inhibit cellular growth to the same extent as in etiolated tissue. Therefore, after horizontal reorientation of light-grown plants, accumulated auxin continues to stimulate cellular elongation so that the stem overshoots 90°, at which point it is then stimulated to curve back in the other direction.

Although ethylene-auxin cross-talk is a major component of the gravitropic response, it is only one of many factors involved, and is not the only mechanism by which ethylene regulates gravitropic curvature. Microarray analysis of horizontally placed light-grown *Arabidopsis* plants reveals that ethylene-responsive transcription factors form a significant functional category of up-regulated genes by 30 min after stimulation. Therefore, horizontal orientation leads to a change in the expression of many ethylene-regulated genes, indicating that ethylene plays a more significant role in gravitropism than previously believed.

7.6 Concluding Remarks

Evolutionarily, it is logical that the gravitropic response for seedling orientation is different than the response of a vegetative stem. In emerging seedlings, gravitropism is part of an adaptive mechanism that allows the plant to move through soil. The seedling stem grows rapidly and responds quickly to changes in orientation until the tissues are exposed to light, when they become less responsive. Seedlings often encounter mechanical perturbation while moving through soil, inducing wound ethylene production that will inhibit elongation and stimulate hook formation to protect the apical meristem. Reducing cell elongation makes the stem thicker and stronger so it can push through the soil. In lighted conditions, vegetative primary stems have reduced sensitivity to ethylene and slower curvature rate than dark-grown stems. For the light-grown plants, phototropism becomes a major factor controlling stem orientation. In these stems, regulation of gravitropism serves to maintain upright orientation and allows the stem to reorient after being knocked to the ground by rain or high winds (lodging). Gravitropism also plays an important role in light-grown *Arabidopsis* inflorescences, where a rapid response may represent an adaptation to quickly raise flower stalks to enhance pollination.

Acknowledgements: We thank Susan Weinstein for her careful reading and helpful advice concerning this manuscript. This research was supported by the National Aeronautics and Space Administration (NAGW-3859), USDA-NRI (2001-35311-11186), and the NASA-EPSCoR West Virginia Space Grant Consortium.

References

Abel S, Nguyen MD, Chow W, Theologis A (1995) *ASC4*, a primary indoleacetic acid-responsive gene encoding 1-aminocyclopropane-1-carboxylate synthase in *Arabidopsis thaliana*. J Biol Chem 270:19093–19099

Abeles FB (1973) Ethylene in plant biology. Academic, San Diego

Abeles FB, Rubinstein B (1964) Regulation of ethylene evolution and leaf abscission by auxin. Plant Physiol 39:963–969

Abeles FB, Morgan PW, Saltveit ME (1992) Ethylene in plant biology, 2nd edn. Academic, San Diego

Bandurski RS, Schultze A, Dayanandan P, Kaufman PB (1984) Response to gravity by *Zea mays* seedlings I. Time course of the response. Plant Physiol 74:284–288

Binder BM, Mortimore LA, Stepanova AN, Ecker JR, Bleecker AB (2004) Short-term growth responses to ethylene in *Arabidopsis* seedlings are *EIN3/EIL1* independent. Plant Physiol 136:2921–2927

Chae HS, Faure F, Kieber JJ (2003) The *eto1*, *eto2*, and *eto3* mutations and cytokinin treatment increase ethylene biosynthesis in *Arabidopsis* by increasing the stability of ACS protein. Plant Cell 15:545–559

Chen G, Rong Y, Li N (2005) *EGY1* encodes a membrane-associated and ATP-independent metalloprotease that is required for chloroplast development. Plant J 41: 364–375

Clifford PE, Reid DM, Pharis RP (1983) Endogenous ethylene does not initiate but may modify geobending – a role for ethylene in autotropism. Plant Cell Environ 6:433–436

Cosgrove DJ (1990) Rapid, bilateral changes in growth rate and curvature during gravitropism of cucumber hypocotyls: implications for mechanism of growth control. Plant Cell Environ 13:227–234

De Grauwe L, Vandenbussche F, Tietz O, Palme K, Van Der Straeten D (2005) Auxin, ethylene and brassinosteroids: tripartite control of growth in the *Arabidopsis* hypocotyl. Plant Cell Physiol 46:827–836

Firn RD, Digby J (1979) A study of the autotropic straightening reaction of a shoot previously curved during geotropism. Plant Cell Environ 2:149–154

Friedman H, Meir S, Rosenberger I, Halevy AH, Kaufman PB, Philosoph-Hadas S (1998) Inhibition of the gravitropic response of snapdragon spikes by the calcium-channel blocker lanthanum chloride. Plant Physiol 118:483–492

Friml J, Wisniewska J, Benkova E, Mendgen K, Palme K (2002) Lateral relocation of auxin efflux regulator PIN3 mediates tropism in *Arabidopsis*. Nature 415:806–809

Folta KM (2004) Green light stimulates early stem elongation, antagonizing light-mediated growth inhibition. Plant Physiol 135:1407–1416

Fukaki H, Fujisawa H, Masao T (1996) Gravitropic response of inflorescence stems in *Arabidopsis thaliana*. Plant Physiol 110:933–943

Golan A, Tepper M, Soudry E, Horwitz BA, Gepstein S (1996) Cytokinin, acting through ethylene, restores gravitropism to *Arabidopsis* seedlings grown under red light. Plant Physiol 112:901–904

Harrison M, Pickard BG (1984) Burst of ethylene upon horizontal placement of tomato seedlings. Plant Physiol 75:1167–1169

Harrison MA, Pickard BG (1986) Evaluation of ethylene as a mediator of gravitropism by tomato hypocotyls. Plant Physiol 80:592–595

Harrison MA, Pickard BG (1989) Auxin asymmetry during gravitropism by tomato hypocotyls. Plant Physiol 89:652–657

Hart JW, MacDonald IR (1981) Phototropism and geotropism in hypocotyls of cress (*Lepidium sativum*). Plant Cell Environ 4:197–201

Kieber JJ (1997) The ethylene response pathway in *Arabidopsis*. Annu Rev Plant Physiol Plant Mol Biol 48:277–296

Kiss JZ, Edelmann RE, Wood PC (1999) Gravitropism of hypocotyls of wild-type and starch-deficient *Arabidopsis* seedlings in space-flight studies. Planta 209:96–103

Lehman A, Black R, Ecker JR (1996) HOOKLESS1, an ethylene response gene, is required for differential cell elongation in the *Arabidopsis* hypocotyl. Cell 85:183–194

Lu BW, Yu HY, Pei LK, Wong MY, Li N (2002) Prolonged exposure to ethylene stimulates the negative gravitropic responses of *Arabidopsis* inflorescence stems and hypocotyls. Funct Plant Biol 29:987–997

MacDonald IR, Hart JW, Gordon DC (1983) Analysis of growth during geotropic curvature in seedling hypocotyls. Plant Cell Environ 6:401–406

Madlung A, Behringer FJ, Lomax TL (1999) Ethylene plays multiple nonprimary roles in modulating the gravitropic response in tomato. Plant Physiol 120:897–906

Masson PH, Tasaka M, Morita MT, Guan C, Chen R, Boonsirichai K (2002) *Arabidopsis thaliana*: a model for the study of root and shoot gravitropism. In: Somerville CR, Meyerowitz EM (eds) The *Arabidopsis* book. Am Soc Plant Biol, Rockville, MD, DOI/10.1199/tab.0043, http://www.aspb.org/publications/arabidopsis

Mattoo AK, Anderson JD, Chalutz E, Lieberman M (1979) Influences of enol ether amino acids, inhibitors of ethylene biosynthesis, on aminoacyl transfer RNA synthesis and protein synthesis. Plant Physiol 64:289–292

Philosoph-Hadas S, Meir S, Rosenberger I, Halevy AH (1996) Regulation of the gravitropic response and ethylene biosynthesis in gravistimulated snapdragon spikes by calcium chelators and ethylene inhibitors. Plant Physiol 110:301–310

Pickard BG (1973) Geotropic response patterns of the *Avena* coleoptile. I. Dependence on angle and duration of stimulation. Can J Bot 51:1003–1021

Pickard BG (1985) Roles of hormones, protons and calcium in geotropism. In: Pharis RP, Reid DM (eds) Encyclopedia of plant physiology, new series, vol 11. Springer, Berlin Heidelberg New York, pp 193–281

Raz V, Ecker J (1999) Regulation of differential growth in the apical hook of *Arabidopsis*. Development 126:3661–3668

Smalle J, Haegman M, Kurepa J, van Montagu M, van der Straeten D (1997) Ethylene can stimulate *Arabidopsis* hypocotyl elongation in the light. Proc Natl Acad Sci USA 94:2756–2761

Stankovic B, Volkmann D, Sack FD (1998) Autotropism, automorphogenesis, and gravity. Physiol Plant 102:328–335

Steed CL, Taylor LK, Harrison MA (2004) Red light regulation of ethylene biosynthesis and gravitropism in etiolated pea stems. Plant Growth Regul 43:117–125

Swarup R, Parry G, Graham N, Allen T, Bennett M (2002) Auxin cross-talk: integration of signalling pathways to control plant development. Plant Mol Biol 49:411–426

Wheeler RM, White RG, Salisbury FB (1986) Gravitropism in higher plant shoots. IV. Further studies on participation of ethylene. Plant Physiol 82:534–542

Wright M, Mousdale DMA, Osborne DJ (1978) Evidence for a gravity-regulated level of endogenous auxin controlling cell elongation and ethylene production during geotropic bending in grass nodes. Biochem Physiol Pflanzen 172:581–596

8 Role of Ethylene in Fruit Ripening

Pravendra Nath, Prabodh K. Trivedi, Vidhu A. Sane, Aniruddha P. Sane

8.1 Introduction

Ethylene, the gaseous plant hormone, has been shown to be involved in almost every phase of plant growth and development, particularly in higher plants (Abeles et al. 1992). This simple two-carbon moiety has the ability to evoke several responses in plants by switching on or off hundreds of genes to affect a process (Mattoo and Suttle 1991). The concerted expression and/or suppression of these genes bring about morphological, physiological, and biochemical changes in plants in diverse events such as seed germination, abscission, fruit ripening, senescence, and various kinds of biotic and abiotic stresses. These genes could be developmentally predetermined and their expression is necessarily mediated by ethylene or other stimuli may evoke response through ethylene. Therefore, genes that express or suppress in presence of ethylene may roughly be termed as ethylene-responsive genes (Giovannoni 2001; Chang and Bleecker 2004).

The role of ethylene in the ripening of fruits was suspected much earlier but it was only in the late sixties and early seventies that experimental evidence began to accumulate. Since then, considerable evidence at the physiological, biochemical, and molecular level has accumulated related to ethylene biosynthesis, its perception by the target cells, the signal transduction cascade involving both positive and negative regulators and, finally, regulation of target gene expression. Most of the information on the role of ethylene in fruit ripening came from studies on tomatoes. Ripening mutants of tomato like *Nr, rin, nor*, etc., have proven very valuable in unraveling how the developmental and ethylene signal is transduced to cause ripening in fruits (Tigchelaar et al. 1978; Giovannoni 2004).

The process of ripening varies in different fruits as far as the involvement of ethylene is concerned. One category of fruits exhibits a burst in ethylene production for a limited period soon after ripening starts and is accompanied by a high rate of respiration (Biale 1964; McMurchie et al. 1972). These fruits are grouped as climacteric fruits, and include banana, mango, apple, tomato, melon, pear, and many other fleshy fruits. The other category of fruits that

---

Plant Gene Expression laboratory, National Botanical Research Institute, Rana Pratap Marg, Lucknow, 226001 India

Ethylene Action in Plants
(ed. by N.A. Khan)
© Springer-Verlag Berlin Heidelberg 2006

does not show this burst of ethylene and CO_2 is classified as non-climacteric fruits, and includes citrus fruits, grapes, strawberries, etc. The role of ethylene in climacteric fruit ripening has been widely studied and several aspects of it are now well established. In contrast, the exact role of ethylene in the ripening of non-climacteric fruit is rather uncertain. Though some aspects of non-climacteric ripening such as rind color differentiation may require ethylene (Chervin et al. 2004; Tesniere et al. 2004), no conclusive evidence is available to assign a characteristic role of ethylene for this category as in the case of climacteric fruits. Hence, in the following pages we will focus on describing the role of ethylene in ripening of the climacteric category of fruits. We have tried to include, in brief, the ethylene biosynthesis, its perception and signal transduction, major genes that are involved in ripening and their regulation, studies on promoters related to ripening, the role of other phytohormones in ripening, and biotechnology related to ripening. Almost every aspect mentioned above has been reviewed individually from time to time; however, there have been constant developments in all these areas to add to our existing knowledge of fruit ripening.

8.2 What is Fruit Ripening?

The term ripening refers to the stage of fruit when it is ready for consumption. In general, ripening may be considered as the changes taking place from the final stage of fruit development through the early stage of its senescence and relates mainly to fleshy fruits (Watada et al. 1984). Biochemically, it can be defined as the summation of changes in tissue metabolism that renders the fruit attractive for consumption by organisms that assist in seed release and dispersal (Adams-Phillips et al. 2004a, 2004b). The stages of ripening have been defined for each individual fruit. For example, in tomato, different stages (mature green, breaker, pink and red) represent the ratio between chlorophyll and lycopene content (Polder et al. 2004). Similarly, the changes in peel color from green to yellow may be used to define the various ripening stages in banana fruit (von Loesecke 1950). In fruits such as apple and melon, which normally ripen on the tree or vine, the stages of ripening are defined on the basis of days post anthesis or fertilization. In non-climacteric fruits, the process of ripening can also be visually distinguished, e.g., in strawberry and lemon, changes in fruit color are indicative of the stage of ripening.

The respiratory climacteric and the burst in ethylene production in climacteric fruits is an important stage in the initiation of ripening (Lelievre et al. 1997). The ethylene signal generated during this period triggers several changes that lead to conversion of starch into free sugars, changes in pH, development of aroma, degradation of chlorophyll, synthesis of carotenoids and flavonoids and pulp and peel softening (Gray et al. 1992; Seymour et al. 1993). Since the majority of changes are strongly correlated with the timing of respiratory climacteric, ripening in these fruits is also subdivided into three major stages: pre-climacteric, climacteric, and post-climacteric. These can be summarized as shown in Fig. 8.1. The changes in various parameters in Fig. 8.1 are only representative and do not conform to any particular climacteric fruit. The nature of various curves will vary depending on the type of fruit, though the general pattern may remain more or less similar in most fruits. The changes in the different ripening parameters such as respiratory climacteric and ethylene production will also depend on whether the fruit has been ripening naturally or artificially by exposure to ethylene or other hydrocarbons. Artificial ripening of climacteric fruits generally hastens the process of ripening. Though it leads to faster spoilage as well, it is practiced largely due to its utility towards early marketing benefits.

Once ripening has been initiated by ethylene, it is an irreversible process and continues uncontrolled. The rapid degradative processes soon make the fruit lose its aesthetic appearance. The initial changes in ripening are brought about by genes that are strongly ethylene responsive. These may activate other genes and other pathways, leading to rapid biochemical changes. These changes probably activate other sets of genes, not necessarily ethylene responsive, that are responsible for progression of ripening followed by genes that finally lead to senescence and spoilage of fruit (Grierson 1987). Since spoilage of fruit is one of the main causes of post-harvest loss of produce, several ways have been devised to prevent these losses. With the advent of biotechnology, it became gradually apparent that genes that are regulated by ethylene could be manipulated through recombinant DNA technology so as to prolong the shelf life of fruits.

In order to achieve this it is important to dissect the ripening process step by step. How a ripening fruit achieves a burst in ethylene production and how

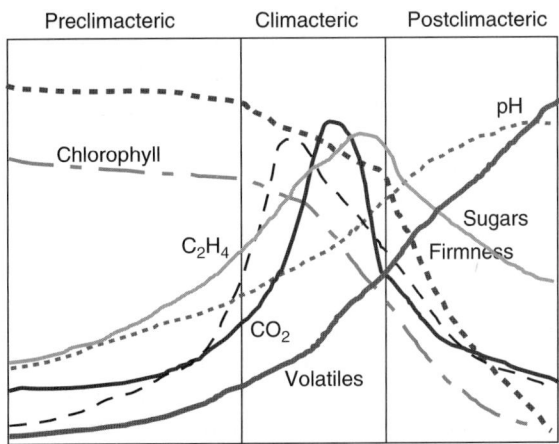

Fig. 8.1. Changes in various parameters during climacteric fruit ripening

the ethylene signal moves towards the target gene is one of the most important aspects in fruit ripening. Simultaneous up and down regulation of several hundred genes during ripening as characterized through microarray analyses indicate that an ethylene-responsive element must be present in promoters of several genes. Strikingly, no such functional common element has yet been determined, making the process of ripening very complicated but interesting. Finally, the kind of success that has been achieved in controlling the process of ripening through recombinant technology is currently more of a scientific success than a commercial one. We shall be discussing some of these aspects in the following sections.

8.3 Ethylene Biosynthesis in Fruits

Ethylene production in fruits as in other plant tissues results from methionine through the Yang cycle (Yang and Hoffman 1984; Bleecker and Kende 2000). S-Adenosylmethionine (S-AdoMet or SAM) is the precursor for ethylene biosynthesis and is synthesized from methionine by SAM synthetase. The rate-limiting step in the ethylene biosynthetic pathway is the conversion of S-AdoMet to 1-aminocyclopropane-1-carboxylate (ACC) by the enzyme ACC synthase, ACS, (Kende 1993). ACC is then oxidized to ethylene, CO_2 and HCN by ACC oxidase (ACO). A salvage pathway works simultaneously to maintain the methionine pool for continuous ethylene production. This salvage pathway works through 5′-methylthioadenosine (MET), a by-product formed by the ACS catalyzed reaction. MET is then converted to methionine.

Small multi-gene families code the ACS and ACO enzymes involved in ethylene biosynthesis. Their expression is differentially governed by various developmental, environmental, and hormonal cues. Ethylene biosynthesis is under positive and negative feedback regulation. In ripening fruits, positive feedback regulation of ethylene biosynthesis is more pronounced and occurs due to activation of the ACS and ACO enzymes leading to autocatalytic synthesis of ethylene. This section will deal with the identification of genes responsible for ethylene synthesis in various fruits and their regulation.

8.3.1 ACC Synthase in Fruits

Most of the studies related to ethylene biosynthesis in fruits are based on tomato, which is a climacteric fruit. Besides tomato, various ACS genes have been identified in different fruits of climacteric and non-climacteric categories like apple (Dong et al. 1992), melon (Yamamoto et al. 1995), pear (Lelievre et al. 1997), cucumber (Shiomi et al. 1998), banana (Liu et al. 1998; Pathak et al. 2003), passion fruit (Mita et al. 1998), citrus (Wong et al. 1999), papaya, etc. So far, eight ACS genes (*LeACS1A, LeACS1B* and *LeACS2–7*) have been identified in tomato (Oetiker et al. 1997; Nakatsuka et al. 1998). Of these, *LeACS2* and *LeACS4* are highly expressed during tomato fruit ripening whereas *LeACS1A* and *LeACS6* express in tomato fruit prior to commencement of ripening. Tomato mutants *Never ripe* (*Nr*) and *ripening inhibitor* (*rin*) were used to identify the ACS genes that are ethylene regulated. These mutant studies have shown that only *LeACS2* was ethylene regulated whereas all others were unaffected by ethylene (Barry et al. 2000). It has been proposed that *LeACS1A* and *LeACS6* are involved in ethylene production in green fruits (system I) but at the time of transition, i.e., unripe to ripe stage *LeACS1A* expression increases and *LeACS4* is induced. This is the time when system II ethylene production starts and is maintained by positive feedback regulation of *LeACS2*. *MaACS1* from banana is related to ripening in banana and its transcript and ACC content increases dramatically at the onset of ripening. It has been suggested that ethylene production in banana fruit is regulated by *MaACS1* until climacteric rise (Liu et al. 1998). In passion fruit, *Pe-ACS1* transcript levels increased during progression of ripening corresponding to the amount of ethylene produced. The other gene *Pe-ACS2* was responsible for low levels of ethylene production during preclimacteric stage (Mita et al. 1998). In *Actinidia chinensis* fruit (a diploid relative of kiwi fruit), ACS mRNA levels increased during climacteric ethylene production but ACS itself was not affected by exogenous ethylene (Whittaker et al. 1997).

Besides being transcriptionally regulated, ACS genes are also regulated post-translationally. Phosphorylation of ACS protein results in increased enzyme activity. Spanu and coworkers (1994) proposed that phosphorylation controls the rate of enzyme turnover rather than regulating specific activity of the enzyme. Various studies carried out on tomato ACS proteins suggest that phosphorylation protects the ACS protein from degradation, which in turn results in increased ACS activity and ACC accumulation and higher ethylene production (Cosgrove et al. 2000; Tatsuki and Mori 2001).

8.3.2 ACC Oxidase in Fruits

The second important enzyme in the ethylene biosynthetic pathway is ACC oxidase. ACO activity increases in preclimacteric fruits before the rise in ACS activity (Lui et al. 1985). ACO, like ACS, is also encoded by a multigene family in various plants (Kende 1993). ACO transcripts have been found in different fruits like tomato (Barry et al. 1996; Nakatsuka et al. 1998), apple (Ross et al. 1992), melon (Bouquin et al. 1997), kiwi (Whittaker et al. 1997), pear, passion fruit (Mita et al. 1998), banana (Liu et al. 1998), etc. These genes have different temporal and spatial expression. In tomato, of the three ACO genes identified, *LeACO1* expresses mainly in ripening fruit. *LeACO2* expression is restricted to anther and *LeACO3* mRNA is present in floral organs with very weak expression in fruits at the breaker stage. Studies on banana ACO genes indicate that *MaACO1* mRNA accumulation is enhanced when ripening commences but its activity decreases rapidly at the late ripening stage due to limitation of its co-factor Fe and ascorbate (Liu et al. 1998). In *A. chinensis*, expression of ACO as well as SAM synthetase increased upon exposure to exogenous ethylene as well as before the climacteric rise in ethylene synthesis. In melon, *Cm-ACO1* expression was induced rapidly by ethylene and wounding (Bouquin et al. 1997) whereas *CmACO3* expressed only in flowers. 1-MCP, which is an ethylene action inhibitor, inhibited *Cm-ACO1* transcript accumulation but no reduction in mRNA abundance was seen in wounded leaves pre-treated with 1-MCP. It was suggested that melon *ACO1* is regulated differently during wounding or ethylene treatment and probably the regulation is through two independent pathways.

8.4 Ethylene Perception and Signal Transduction in Fruits

The ethylene-signaling pathway has been studied in great detail in *Arabidopsis* but it has still not been fully elucidated. The new information coming through various *Arabidopsis* mutants and genetic studies have opened new frontiers in the understanding of the signal perception and its transduction. The ethylene signal transduction pathway in *Arabidopsis* has been reviewed extensively by several groups (Bleecker and Kende 2000; Schaller and Kieber 2002; Chang and Bleecker 2004; Chen et al. 2005). The pathway for ethylene signaling in fruits shows conservation of some components of basic signaling pathway to those described in *Arabidopsis* although several genes in the pathway are yet to be identified in fruits (Klee 2004).

A family of five membrane-localized receptors, which show homology to bacterial two-component system, perceives ethylene in *Arabidopsis*. These receptors are negative regulators of ethylene response. The next component of signal transduction pathway is CTR1, which shows homology to Raf family (Ser/ Thr kinase) of mitogen-activated protein kinase kinase kinase (MAPKKK). CTR1 is also a negative regulator of signal transduction pathway. Since CTR1 is a MAPKKK, there are speculations that a MAP kinase cascade may operate in ethylene signaling (Ouaked et al. 2003). However, the components immediately down stream of CTR1 have neither been identified in *Arabidopsis* nor in fruits. Further downstream lies EIN2, which is a membrane-bound Nramp metal transporter. It is an essential positive regulator of signaling although its mode of action is not clearly understood. Transcription factors like EIN3 and ERF1 (ethylene response factors) come at the end of signaling pathway and are responsible for activation of genes that actually bring about responses to a given stimulus. This family of trans-acting factors is localized in the nucleus (Chao et al. 1997; Solano et al. 1998). In *Arabidopsis*, EIN3 is regulated post-translationally by ethylene through proteasome-dependent proteolysis (Guo and Ecker 2003). Homodimers of the EIN3 family bind to ethylene response factor (ERF1), which belongs to a large multigene family of transcription factors. According to the most acceptable model of ethylene signal transduction, ethylene receptors, in absence of ethylene, are in active form and signal CTR to block the downstream components. Binding of ethylene to receptors inactivates them and releases negative regulation of CTR. As a result, EIN2 is activated and induces transcription of EIN3/EIL transcription factors. This allows ethylene responses to occur (reviewed by Chang and Stadler 2001; Guo and Ecker 2004).

In ripening tomato fruits, six putative ethylene receptors LeETR1–6 have been identified (Wilkinson et al. 1995; Payton et al. 1996; Zhou et al. 1996; Lashbrook et al. 1998b; Ciardi and Klee 2001). LeETR3 (originally and historically known as Nr) was the first ethylene receptor that was identified in tomato through the isolation of *Never-ripe (Nr)* gene from fruit ripening locus. Nr lacks the receptor component and is structurally similar to ERS1 of *Arabidopsis*. All these tomato receptors are expressed in various temporal and spatial patterns. *LeETR1* and *LeETR2* are present constitutively through out the development. *Nr* and *LeETR4* are the most abundant transcripts during ripening. These two, however, also show expression during abscission, senescence, and pathogen attack. *LeETR5* is expressed only in flowers, fruits, and during biotic stress such as pathogen infection. Tomato ethylene receptors are functionally redundant in nature like their counterparts in *Arabidopsis* as the repression of individual genes except for *LeETR4*, did not affect the ethylene sensitivity in transgenic plants (Tieman et al. 2000). Even Nr, which in a mutated form inhibits ripening, is not required for ripening since reduced levels of Nr do not affect ripening. Transgenic lines with down regulation of *Nr (LeETR3)* show instead a functional compensation through increased levels of *LeETR4*. In contrast, *LeETR4* knockouts do not induce *Nr* transcripts. Different genetic studies indicate that *LeETR4* is the most important receptor for fruit ripening as well as some other processes. Unlike in *Arabidopsis*, where a minimum of three receptors are required to be knocked out for ethylene sensitivity (Hua and Meyrowitz 1998), *LeETR4* suppression in tomato results in an ethylene hypersensitive phenotype and transgenic fruits show early ripening. It has been hypothesized that *LeETR4* monitors the receptor levels and initiates *de novo* synthesis of new receptors depending on the ethylene response. Besides tomato, ethylene receptors have also been identified in other fruits. Three receptors, *PcETR1a, PcERS1a* and *PcETR5* have been identified in pears (El-Sharkawy et al. 2003), two in musk melon (Sato-Nara et al. 1999) and one each in passion fruit (Mita et al. 1998) and peach (Rasori et al. 2002). All these receptors from different fruits are highly conserved and show homology to bacterial two-component system. These receptors exhibit differential expression during developmental phase and some of these are induced by ethylene (Chen et al. 2005). Recently, three ethylene receptors were identified in non-climacteric fruit strawberry (Trainotti et al. 2005). Two receptors *FaEtr1* and *FaErs1* belong to type I receptors whereas *FaEtr2* belongs to type II group of receptors. Type II receptors have degenerate histidine kinase domain and have weaker affinity for CTR1 (Cancel and Larsen 2002). All these receptors have increased expression during fruit development from green to white stage. *FaEtr2* is most abundant receptor during ripening phase of strawberry. Increased expression of ethylene receptors during ripening in non-climacteric fruits suggests that ethylene, though present in very small amount, may be sufficient to trigger some aspects of ripening in these fruits.

There is not much information available on the downstream components of the signaling pathway in fruits. Unlike in *Arabidopsis* where only one *CTR1* gene has been identified, at least four *CTR1* homologues have been identified in tomato. One of these, *LeCTR1*, has been shown to functionally complement *Arabidopsis ctr1* mutation (Leclercq et al. 2002). Different studies on CTR homologues from tomato indicate that these genes express at higher levels in ripening fruits as compared to unripe fruits (Adams-Phillips et al. 2004). This result is surprising, as CTR1 from *Arabidopsis* has been shown to be a negative regulator of ethylene responses. Virus induced gene silencing (VIGS) of *LeCTR1* gene resulted in constitutive ethylene phenotypes in tomato confirming the role of *LeCTR1* as negative regulator of ethylene responses in tomato (Liu et al. 2002). Whether multiple CTR from tomato bind to specific receptors is yet not known. It has been suggested that the transcriptional increase in receptor and CTR1 genes may help in increasing receptor molecules and serve as a mechanism to restore ethylene sensitivity and responsiveness under various conditions. A CTR homologue has also been identified from pear (El-Sharkawy et al. 2003). According to a model as reviewed by Klee (2004) and Adams-Phillips et al. (2004) ethylene binds to receptors to inactivate them and in turn inactivate CTR. The ratio of receptors and CTR encoded by different family members (and for different tissues and responses) might represent a mechanism to optimize ethylene responses in tomato and probably in other fruits.

Of the other downstream components, only one EIN2 homologue from tomato (*LeEIN2*) has been identified and found to be ethylene non-inducible with constitutive expression throughout fruit development. Homologues of EIN3 and ERFs have also been identified in tomato and some other fruits and are discussed later in the section.

8.5 Gene Expression and Regulation during Ripening in Fruits

The transition from fruit development to fruit ripening is marked by large-scale changes in gene expression. In this section, we will discuss the expression and regulation of some of the major classes of genes during ripening. The ripening program seems to consist of a developmental trigger that is responsible for the initiation of ethylene synthesis and other changes and an ethylene signal cascade that triggers the expression of several ripening-related genes. Based on the various biochemical changes that take place during ripening, the changes in gene expression can be categorized as those related to (i) developmental cues (ii) activation of transcription factors, (iii) cell wall modification/hydrolysis (iv) pigment and volatile synthesis and (v) other ripening-related genes isolated by differential screening.

8.5.1 Developmental Regulation

Studies on tomato mutants such as *rin*, *nor*, *Cnr* have greatly facilitated the understanding of ripening. While *rin* and *nor* are non-ripening recessive mutants, *Cnr* is a dominant colorless non-ripening mutant. In these mutants, ripening is inhibited and there is no respiratory climacteric and no autocatalytic ethylene synthesis in fruits, indicating a block upstream of the ethylene cascade. However, responsiveness to ethylene is not affected as evident from the expression of several known ethylene-responsive genes such as *E4*, *E8*, *PG*, etc in mutant fruits upon treatment with exogenous ethylene. Nevertheless, exogenous ethylene is unable to initiate ripening in these fruits, indicating the presence of another ethylene independent pathway that needs to be active and runs parallel to the ethylene-mediated processes. The identity of the affected *rin* gene, *LeMADS-RIN*, as a MADS box protein gene revealed the role of these proteins also in ripening (Vrebalov et al. 2002). These genes are known to play an important role in several plant developmental processes especially in flower development. Moreover, these genes function as heterodimers or higher-order multimers of other transcription factors, indicating that other MADS box proteins or transcription factors may also be involved in determining the developmental timing of ripening. Reports from Giovannoni's lab at Cornell indicate that the *Nor* gene is also a transcription factor that shows no homology to other MADS box transcription factors (Adams-Phillips et al. 2004). *LeMADS-RIN* is expressed in mature green fruits and is not ethylene responsive. Interestingly, homologues of this gene are also expressed in strawberry, a non-climacteric fruit, also at the mature green stage prior to ripening, indicating that the developmental program for ripening may be common to both climacteric and non-climacteric fruits. Further studies on repression of the *LeMADS-RIN* homologue in strawberry and micro-array analyses of tomato genes affected in RIN over-expressing and under-expressing plants may shed light on the down stream targets of these genes in ripening.

8.5.2 Ethylene Regulation of Transcription Factors

Once the developmental pathway initiates the process of ripening, ethylene synthesis and perception lead to large scale changes in gene expression that drive the ripening via the ethylene dependent pathway. The ethylene signal cascade (described earlier) activates transcription factors such as EIN3 (ethylene insensitive) and EIN3-like (EIL) genes and ERFs (ethylene response factor). Several EIN3 and EIN3-like genes have been identified in tomato and other fruits such as melon. At least four EILs, *LeEIl1, LeEIL2, LeEIL3* and *LeEIL4* have been identified in tomato. The first three genes complemented the *ein3-1* mutation in *Arabidopsis*. The tomato LeEIL family of genes is not ethylene inducible with unaltered transcript expression throughout the growth and development. Repression of either of the genes individually does not affect any aspect of fruit development indicating redundancy between the EIN3 like proteins. However, repression of *LeEIL1, 2* and *3* genes simultaneously affects several aspects of plant development including fruit ripening demonstrating their role in ripening (Tieman et al. 2001). Recently, it was shown that constitutive expression of *LeEIL1* in the *Nr* mutant of tomato (a dominant ethylene receptor mutant) partially restored ripening in tomato (Chen et al. 2004a). Expression of a subset of ethylene-responsive genes such as *PG* and *tomloxB* was higher in transgenic EIL1 over-expressing *Nr* plants as compared to *Nr* plants but those of other genes such as *E4* and *LeACO1* was not affected. Moreover, other ethylene responses such as seedling triple response were not restored in *LeEIL1* over-expressing plants. The results showed the involvement of the *LeEIL1* in at least some aspects of ripening. The specific roles of other EILs has however not yet been established. EIN3 is known to activate ERFs in *Arabidopsis* by binding specific *cis* elements. Recent studies have shown that ERFs are also expressed in tomato fruits (Tournier et al. 2003). Four ERFs were identified in tomato of which *LeERF2* was shown to express predominantly in fruits. These ERFs had the ability to bind the GCC box element, a *cis* element present in several defense responsive genes in plants. Further studies on repression and over expression of these ERFs in fruits may provide a better understanding on the role for these ERFs in fruit ripening.

Transcription factors such as the ASR protein (*A*BA *s*tress *r*ipening-related protein) have also been identified during the course of ripening in fruits such as apricot and banana (Mbeguie-A-Mbeguie et al. 1997). The ASR protein is a Zn-containing DNA binding factor that affects the expression of several genes upon over-expression (Kalifa et al. 2004). In addition, recent EST analyses of tomato cDNAs have shown that other transcription factors may also be involved in ripening. At least 18 putative transcription factors were upregulated during ripening while at least 14 were repressed during ripening.

8.5.3 Gene Expression during Softening

Softening of fruits has been one of the most exhaustively studied processes since its uncontrolled progression leads to post-harvest destruction. Different fleshy fruits are characterized by differences in their cell wall composition that necessitates the action of various cell wall hydrolases and cell wall modifying proteins. Two major components in the cell wall include pectins and cellulose/hemicelluloses that form a complex network with matrix glycans and glycoproteins enmeshed in between. Expressions of several cell wall hydrolases such as polygalacturonases, cellulases, pectin methyl esterases, galactosidases, pectate lyases, xyloglucan transglucosylase/hydrolases and expansins have been studied extensively in several fruits (reviewed by Brummell and Harpster 2001). Each of these enzymes shows specificity towards one particular component of the cell wall. The concerted action of all or several of these genes is believed to bring about softening of the fruit pulp. Studies so far reveal that no single gene is entirely responsible for softening. Thus, strategies where one gene has been repressed have not led to many changes in softening. Additional complexities are seen in the form of expression of wall hydrolase inhibitor proteins such as PGIP for PG and PMEI for PME (Irifune et al. 2004). Interestingly, many of the genes exist as a multigene family with several members showing transcriptional up regulation during the onset of ripening and softening with often overlapping patterns of expression. The role of several members of the same gene family in ripening-related expression is not yet clear. However, it is likely that as softening proceeds the release of wall components by hydrolysis may cause a shift in pH of the environment of the cell wall, necessitating the need for proteins that are active at different pH conditions. Multiple expansins have been shown to be active in tomatoes, peach, pear, apricot, apple, banana, etc., with their expression being regulated by ethylene or the ripening stage (Harrison et al. 2001; Hiwasa et al. 2003). *LeExp1* in tomato, *MaExp1* in banana and *MiExpA1*, in mango, are under transcriptional control of ethylene and rapidly accumulate after ethylene treatment (Rose et al. 1997; Trivedi and Nath 2004; Sane et al. 2005). For many other genes, treatment with 1-MCP may block expression but this may be an indirect effect due to inhibition of ripening. Polygalacturonases have also been studied in many fruits and were considered important since in tomato, PG accounts for nearly 2% of the total fruit mRNA. However, subsequent studies show that the activity of PG may vary from fruit to fruit and while it is present in high levels in tomato, avocado, and peach, it may not be very actively expressed in other fruits. In tomato, PG transcription is under control of ethylene with even low levels of ethylene sufficient to induce transcription (Della Penna et al. 1989). In banana, five PG genes were shown to differentially express during ripening (Asif and Nath 2005) although the overall PG activity is low. Pectate lyases have been studied extensively in banana and strawberry where they are expressed to very high levels. In banana, the pectate lyase gene is strongly induced by ethylene. Galactosidases have been shown to be active in tomato, papaya, and strawberry. Enzymes like endo-1,4-α-mannanase and α-xylosidase are also expressed during ripening. Other genes such as XTHs, galactosidases, α-1,3- and α-1,4-glucanases have also been studied in fruits like tomato, papaya, etc. For most of the genes, treatment with 1-MCP may block expression but this may be an indirect effect due to inhibition of ripening. A direct transcriptional control by ethylene has not been conclusively demonstrated and they may instead be regulated by secondary factors that appear after ripening has been initiated by ethylene.

8.5.4 Genes Involved in Changes in Pigments and Volatiles

Changes in pigment, particularly the degradation of chlorophyll and the increase in flavonoids, carotenoids, lycopene, etc., as well as the production of volatiles, are other major changes that take place during ripening and function as strong visual and olfactory signals to attract animals and birds for seed dispersal. Many of these changes are ripening-dependent. Changes related to pigment conversion during ripening are influenced by light with phytochromes playing an important role. The *DET1* gene, a repressor of light signaling, was shown recently to influence levels of carotenoids and flavonoids in tomato (Davuluri et al. 2005). The phytoene synthase gene from tomato, responsible for formation of the carotene phytoene is also regulated by light. Other genes such as lyocopene beta cyclase play a role in lycopene synthesis and the balance between lycopene and carotene levels.

Volatiles also form an important component of several fleshy fruits and aid in attracting animals. They mainly consist of esters, aldehydes, and alcohols and are produced from fatty acids by the action of alcohol acyl CoA transferases, lipoxygenases, and alcohol dehydrogenases. The variety of substrates used by the alcohol acyl transferases may be responsible for the range of volatiles produced in fruits. Production of several volatiles in fruits is governed by ethylene. Treatment with 1-MCP or antisense repression of ethylene production in fruits has been shown to affect volatile production in tomato, apples, mango, melons, and apricot. In tomato, at least three lipoxygenase genes, *tomloxA, tomloxB* and *tomloxC* (Griffiths et al. 1999) express during fruit ripening. Although all three are affected by ethylene, only reduction of *tomloxC* in transgenic plants led to a significant reduction in C6 volatiles (Chen et al. 2004b). In apple, reduction in ethylene reduced ester production but did not affect aldehyde production. The expression of alcohol acyl transferase but not alcohol dehydrogenase was ethylene dependent (Defilippi et al. 2005). In melons, ethylene affected the reduction of fatty acids and aldehydes, resulting in reduction of aliphatic esters whereas alcohol acetylation had both ethylene dependent and independent components (Flores et al. 2002).

8.5.5 Other Ripening-Related Genes Isolated by Differential Screening

Apart from these several studies on fruit ripening have been performed by methods that involve differential screening of ethylene untreated (unripened) and ethylene treated (ripened) fruits in an effort to identify genes expressed during ripening. These studies, performed in tomato (Zegzouti et al. 1999), banana (Clendennen and May 1997), melon (Hadfield et al. 2000) and several other fruits such as strawberry, peaches, pear, avocado have led to identification of ethylene regulated genes such as a *E4* (a methonine sulfoxide reductase protein), *E8* (a dioxygenase involved probably in feedback regulation of ethylene), genes involved in post-transcriptional regulation such as *ER24* (similar to transcriptional co-activator MBF1) and *ER49* (translational elongation factor EF-Ts), *ER50* (a RNA helicase) and *LeRab11a* (a GTPase involved in trafficking of cell wall modifying enzymes). Ripening in many fruits, both climacteric as well as non-climacteric, such as banana, melon, avocado, cherries, and raspberry also induces genes that are expressed during stress/defense responses. These include chitinases, α-1,3-glucanases, thaumatin like proteins, isoflavone reductase like genes and lectins and probably expressed as a response to ethylene release (Clendennen and May 1997). Since ripening involves conversion of starch and acids to sugars, genes related to carbohydrate metabolism such as sucrose synthase, invertase, sucrose phosphate synthase, malate synthase, PEP carboxykinase are also differentially expressed during ripening (Bahrami et al. 2001; Fung et al. 2003; Pua et al. 2003).

Some genes related to mitochondrial function such as alternative oxidase have been shown to express during the ripening in apple and mango (Considine et al. 2001). This may not be surprising given the large respiratory climacteric in these fruits. In some fruits, this climacteric is associated with an increase in temperature that probably results from the action of the alternative oxidase or the uncoupler protein. In recent years, large-scale genome sequencing and generation of EST databases of fruit cDNAs has made microarray analyses of genes expressing in fruits possible. These analyses have led to information on expression of hundreds of genes during different stages of fruit ripening. However, understanding the specific function of each gene in ripening will require more detailed studies.

8.6 Role of Other Hormones and Metabolites during Ripening

Most developmental and adaptive processes in plants result from a complex interaction between two or more hormones. Studies on *Arabidopsis* mutants indicate that a single hormone can regulate an amazingly diverse array of cellular and developmental processes, while at the same time multiple hormones often influence a single process (Gray 2004). The extent of cross talk and the type of interactions (synergistic or opposing/antagonistic) vary from process to process. Ripening in fruits is one such process where besides ethylene other hormones and biochemicals also participate either positively or negatively. These hormones could influence the developmental switch that leads to initiation of ripening under normal conditions or even influence the timing of ripening under unfavorable environmental conditions. There are evidences of a possible cross talk amongst various phytohormones. However, the molecular basis of signal integration for a coordinated regulation between these hormones during ripening is still unclear. In this section, we shall discuss three hormones: auxin, jasmonic acid, and abscisic acid, whose role in fruit ripening is suspected. We shall also briefly address the role of sugar, polyamines, and salicylic during fruit ripening.

8.6.1 Auxin

Auxins in general are known to be involved in fruit development but inhibit ripening. Fruits show a changing capacity for auxin conjugation as the ripening process proceeds not only in tomato but also in many other fruits tested, and there is a greatly diminished capacity to form indole-3-acetyl-aspartate (Slovin and Cohen 1993). These findings show that fundamental changes in the ability of fruit tissue to regulate auxin levels by conjugation occur during ripening. Though auxins have mostly been related to fruit growth and development rather than ripening, treatment of grape (*Vitis vinifera* L.) berries with the synthetic auxin like compound benzothiazole-2-oxyacetic acid (BTOA) caused a delay in the onset of ripening of approximately 2 weeks (Davies et al. 1997) whereas pretreatment of banana fruit with auxin accelerated expression of ripening specific expansin (Trivedi and Nath 2004). However, pretreatment of banana with IAA inhibited activities of major cell wall hydrolases (Lohani et al. 2004). This shows variable effects of auxin during ripening. Auxin response during root hair development, hypocotyl development, and other cell growth and developmental processes has been reported to intertwine with ethylene response. However, a direct involvement of auxin during ripening is not very clear. Auxin has been reported to act opposite to ethylene response in several processes in a dose-dependant manner, but its role during fruit growth and development sometimes extends during the early phase of ripening and is suspected to relate to sensitivity of ethylene response. It is interesting to note that several genes, which are related to ripening, are controlled by auxin during fruit growth and development. For example, expression of *LeExp2*, *LeExt1*, and *Cel7* was undetectable or negligible at the onset and during the course of fruit ripening but was detected during fruit growth and regulated by auxin. This shows its specific role in regulating cell wall loosening during fruit growth but not in ripening-associated cell wall disassembly (Catala et al. 2000). A GH-3-like gene, *CcGH3*, which is regulated by auxin, has been shown to express during fruit ripening in *Capsicum chinense*. When this gene was over expressed in tomato, it hastened ripening of ethylene-treated fruit (Liu et al. 2005). An auxin-inducible ACS has also been reported during pre-climacteric stage of ripening in melon (Ishiki et al. 2000).

8.6.2 Jasmonate and Jasmonic Acid

Saniewski et al. (1987) noticed that exogenous application of jasmonate promoted climacteric fruit ripening by increasing ethylene production in apples. It was also noticed that it accelerated α-carotene production and chlorophyll degradation in Golden Delicious apple peel (Saniewski and Czapski 1983; Perez et al. 1993), an essential feature of fruit ripening. Fan et al. (1998) demonstrated that ripening-related aroma compound production was accelerated in apple upon application of jasmonate. However, a conclusive role for endogenous jasmonate in regulating climacteric fruit ripening has still not been demonstrated. From the above studies, it was concluded that transient increase in jasmonate concentrations in apple and tomato might occur during the onset of fruit ripening and that jasmonates are involved in early steps in the modulation of climacteric fruit ripening. It was also found that activities of both ACS and ACO were stimulated in the concentration range of 1–100 µM of jasmonate, however continuous exposure to 1,000 µM inhibited both ACS and ACO activities. The convergence of ethylene and jasmonate pathways at the transcriptional activation level of ERF1 has been shown by Lorenzo et al. (2003). The activation of ERF1, which encodes a transcription factor that regulates the expression of pathogen response genes, can be obtained rapidly by ethylene or jasmonate and synergistically by both hormones. In addition, both signaling pathways are required simultaneously to activate ERF1 expression because mutations that block either of them prevent ERF1 induction by both the hormones either alone or in combination. Further, 35S:ERF1 expression can rescue the defense response defects in *coi1* (*coronatine insensitive*) and *ein2* (*ethylene insensitive2*) mutants; therefore, it is a likely downstream component of both ethylene and jasmonate signaling pathway. Various aspects of the modulation of fruit ripening by jasmonates have been reviewed by Creelman and Mullet (1997).

8.6.3 Abscisic Acid

That abscisic acid enhances fruit ripening of both climacteric and non-climacteric category was known earlier (Parikh et al. 1986; Brady 1987) and it was suspected that various ripening parameters might be related to higher ethylene production. However, the mechanism by which ABA stimulates ethylene production is still not clear. It was postulated that ABA might have a direct effect on ethylene biosynthesis through enhanced production of ACC synthase (Goren et al. 1993). Jiang et al. (2000) reported that induction of ripening in banana fruit by ABA was partially inhibited by pretreatment with 1-MCP indicating that ABA action was mediated by ethylene. This was further elaborated by Lohani et al. (2004) who showed that ABA treatment of banana resulted in higher activities of cell wall hydrolases and softening in the presence of ethylene, and these activities were suppressed significantly by 1-MCP treatment. It is therefore possible that ABA may exert its effect by increasing the sensitivity of the fruit towards ethylene. Transcripts of an abscisic acid responsive transcription factor gene, *Asr1*, were reported to accumulate during melon fruit development particularly at the RP stage and did not show its expression in other tissues (Hong et al. 2002). *Asr* gene expression was earlier reported to be activated during several stresses as well as during fruit ripening and abscisic acid treatment (Gilad et al. 1997). However, the genes activated by *Asr* during fruit ripening and after ABA treatment are not yet known and thus the coordination between the ABA and ethylene through Asr is still unclear.

8.6.4 Other Metabolites

Amongst other metabolites, **salicylic acid** is known to affect ripening. Exogenous applications of SA to apple, peach, banana, mango, and tomato fruits have been reported to delay ripening in these fruits. Leslie and Romani (1988) demonstrated an inhibitory effect of SA on the conversion of ACC to ethylene due to suppression of ACC oxidase activity. Li et al. (1992) demonstrated an inhibitory effect of SA on the accumulation of wound induced ACC synthase transcript. Ding and Wang (2003) showed that ripening process in mature green tomato was enhanced by 0.1 mM methyl salicylate and by 0.01 mM during breaker stage. But in fruit at turning stage even 0.01 mM SA inhibited the ripening process. High concentration (0.5 mM) prevented red color development, ethylene production and respiration rate in all maturity stages. They concluded that increased ethylene production was mediated by depressing the negative feed back regulation of *LeACS6* gene and increasing the expression of *LeACS2* and *LeACS4* through positive feedback. Recently, Leclercq et al. (2005) characterized an isoform of calcium dependant protein kinase(CDPK) from tomato that was induced by ethylene and salicylic acid. *LeCRK1*, showed significant accumulation of its transcript during fruit ripening and was also detected in stem, leaf, and flower tissue. It was noteworthy that *LeCRK1* transcript was undetectable in natural tomato ripening mutants such as *Nr, rin* and *nor*. Though salicylic acid is known to have an antagonistic effect on ethylene, recent developments indicate that it might have synergistic effect in a dose dependant manner in certain cases.

Polyamines are metabolites that utilize a common precursor of ethylene biosynthesis, S-adenosyl methionine. It was believed that polyamine biosynthesis would negatively affect ethylene biosynthesis due to competition for a common substrate and therefore inhibit ripening. It was shown that direct application of putrescine, spermidine and spermine might inhibit ACC synthase activity in tomato (Li et al. 1992). Similar results were obtained for avocado fruit. However enhancing polyamine biosynthesis in tomato did not significantly affect the ripening process (Mehta et al. 2002).

Modulation of ethylene response and signaling process by *sugars* has been reviewed by Sinha et al. (2002). Recent molecular analyses have revealed direct, extensive glucose control of abscisic acid biosynthesis and signaling genes that partially antagonizes ethylene signaling during seedling development under light (Leon and Sheen 2003). Yanagisawa et al. (2003) showed that glucose enhances degradation of EIN3, a key transcriptional regulator in ethylene signaling, through plant glucose sensor hexokinase. Ethylene, by contrast, enhances the stability of EIN3. There are also indications that sugar may exert its effect at the level of CTR and MAP kinases. However, exact role of sugar in controlling ripening is not known. Several studies with sugar and ethylene mutants in *Arabidopsis* show differential responses in presence of ethylene and sugar respectively suggesting possibility of a strong cross talk between sugar and ethylene at down stream level of ethylene signal. It is likely that sugar may play important role at transcriptional regulation level and regulate process of ripening.

8.7 Biotechnological Usage of Ethylene Biology

As described previously, post-harvest losses in fruit are high and need to be controlled. Biotechnology provides a promise for reducing post-harvest losses in fruit by manipulation of the ripening process in fruits. The knowledge generated by gene expression studies related to ethylene biosynthesis and perception during fruit ripening and during the progression of ripening and softening has been utilized to modify fruits for higher post-harvest life and better nutritional qualities. Expression of sense, antisense or RNAi constructs of the different genes has allowed different groups to modify different pathways that affect fruit. In this section, main emphasis has been given to achievements made in the area of delayed ripening and softening (Table 8.1).

8.7.1 Transgenic Fruits with Altered Ethylene Production

As previously discussed, ethylene biosynthesis starts from SAM. The enzyme SAM decarboxylase, converts SAM to decarboxylated SAM, which is then utilized for polyamine biosynthesis. This gene has been isolated from different plants and has been utilized to modify fruits with an idea that overexpression of *SAMDC*, might enhance the flux of SAM through the polyamine pathway, thus reducing the amount available for ethylene biosynthesis. Tomato plants transformed with yeast *SAMDC* gene under control of E8 promoter exhibited enhanced fruit lycopene content, better fruit juice quality, and vine life. The rate of ethylene production in transgenic fruits was higher than in the non-transgenic control fruit, suggesting that polyamine and ethylene biosynthesis pathways may act simultaneously in ripening tomato fruit (Mehta et al. 2002). Another enzyme that hydrolyses SAM, SAM hydrolase, found only in bacteriophage T3, has also been utilized to alter ethylene level in fruits. The use of E8 promoter with SAM hydrolase in transgenic tomato plants, reduced ethylene levels in fruits as compared to controls (Good et al. 1994).

In tomato, the entire ACS gene with its untranslated region was used to develop transgenic plants. Transgenic tomatoes showed 99.5% decrease in

Table 8.1. Transgenic plants with altered fruit ripening

Gene	Promoter	Plant transformed	Consequences	References
(A) Altered ethylene biosynthesis				
Yeast *SAMDC*	Tomato E8	Tomato	Enhanced fruit phytonutrient content, fruit juice quality and vine life, higher rate of ethylene production	Mehta et al. (2002)
Bacteriphage SAM hydrolase	Tomato E8	Tomato	Reduced ethylene production	Good et al. (1994)
Tomato ACS gene	CaMV 35S	Tomato	Decreased ethylene production, fruits do not ripen	Oeller et al. (1991)
Tomato ACS	CaMV 35S	Tomato	30% decrease in ethylene production by fruits, increased shelf life	Liu et al. (1998)
Apple ACS/ACO	CaMV 35S	Apple	Reduced ethylene production, firmer fruits with an increased shelf-life	Dandekari et al. (2004)
P. chloraphis ACC deaminase	CaMV 35S	Tomato	Delayed fruit ripening	Klee et al. (1991)
Tomato ACO	CaMV 35S	Tomato	Decrease in ethylene production	Hamilton et al. (1990)
Melon ACO	CaMV 35S	Melon	Reduced ethylene production, rind stayed greener during ripening, increased concentration of sucrose and citric acid	Flores et al. (2001)
Tomato ACO RNAi	Modified CaMV 35S	Tomato	Decreased ethylene production, delayed fruit ripening, prolonged shelf life	Xiong et al. (2005)
(B) Altered ethylene perception				
Arabidopsis etr1-1	CaMV 35S	Tomato	Decreased ethylene perception, delay in fruit ripening	Wilkinson et al. (1997)
Tomato ETR1 (receiver domain and 3'-UTR)	Enhanced CaMV 35S	Tomato	Reduced LeETR1 transcript levels, delayed abscission, no effect on softening	Whitelaw et al. (2002)
Tomato NR	CaMV 35S	Tomato	Less sensitive to ethylene	Ciardi et al. (2000)
Tomato *ETR4*	CaMV 35S	Tomato	Loss of flowers and early fruit ripening, more sensitive to ethylene	Tieman et al. (2000)

Gene	Promoter	Species	Phenotype	Reference
Tomato NR	CaMV 35S	Tomato	Fruits ripen normally	Hackett et al. (2000)
Tomato EIL1	CaMV 35S	Tomato Nr background	Restoration of ripening, color change	Tieman et al. (2001)
(C) Altered fruit softening				
Tomato PG	CaMV 35S	Tomato	Suppression of PG mRNA and protein accumulation, no effect on fruit softening	Sheehy et al. (1988); Smith et al. (1988)
Tomato PG	Tomato E8	Tomato rin background	Increase in PG activity, no change in softening	Giovannoni et al. (1989)
Tomato PME	CaMV 35S	Tomato	Suppression of PME mRNA and protein accumulation, no effect on fruit ripening and softening	Tieman et al. (1992)
Tomato TBG1	CaMV 35S	Tomato	Suppression of exo-galactanase mRNA but total activity remains unaffected, no effect on fruit softening	Carey et al. (2001)
Tomato TBG4	CaMV 35S	Tomato	Fruit firmness higher than control fruit	Smith et al. (2002)
Tomato TBG6	CaMV 35S	Tomato	Increased fruit cracking, reduced locular space, doubling in the thickness of the fruit cuticle, reduction in fruit firmness	Moctezuma et al. (2003)
Strawberry PL	Enhanced CaMV 35S	Strawberry	Significantly firmer fruits	Jiménez-Bermúdez et al. (2002)
Tomato Cel1 and Cel2	CaMV 35S	Tomato	Decrease in mRNA accumulation, no effect on fruit softening	Lashbrook et al. (1998); Brummell et al. (1999a)
Bell pepper Cel1	CaMV 35S	Bell pepper	Reduced immunodetectable CaCell protein, no effect on fruit softening	Harpster et al. (2002)
Tomato Exp1 (antisense)	CaMV 35S	Tomato	Reduced fruit softening during ripening	Brummell et al. (1999b)
Tomato Exp1 (sense)	CaMV 35S	Tomato	Increased fruit softening	Brummell et al. (1999b)
Tomato PG X Tomato Exp1	CaMV 35S	Tomato	Significantly firmer fruits, juice more viscous	Powell et al. (2003)
Tomato DHS	CaMV 35S	Tomato	Normal fruit ripening but delayed postharvest softening	Wang et al. (2005)

ethylene production and did not ripen without exogenous treatment of ethylene (Oeller et al. 1991). In another attempt, anti-ACS containing transgenic tomato plants showed a 30% decrease in ethylene production by fruits. The shelf life of transgenic tomato fruits was at least 60 days at room temperature without significant change in hardiness and color. After 15–20 days of treatment of the transgenic fruits with ethylene, most of the tomatoes reached the ripe stage (Liu et al. 1998). Apple fruit obtained from plants silenced for ACS expectedly showed reduced autocatalytic ethylene production (Dandekari et al. 2004). Ethylene-suppressed apple fruits were significantly firmer than controls and displayed an increased shelf life. No significant difference was observed in sugar or acid accumulation suggesting that sugar and acid accumulation is not directly under ethylene control. Interestingly, a significant and dramatic suppression of the synthesis of volatile esters was observed in fruit silenced for ethylene.

ACC deaminase, which is present in soil microorganisms and converts ACC to ammonia and α-ketobutyrate, has also been used to delay ripening in tomato fruits. Overexpression of ACC deaminase under control of CaMV 35S promoter produced tomato fruits with delayed fruit ripening (Klee et al. 1991).

Antisense transgenic lines of different plants have also been raised with anti-ACO gene to alter ethylene biosynthesis. Transgenic tomato carrying antisense ACO gene showed decrease in ethylene production (Hamilton et al. 1990). Ayub et al. (1996) produced a transgenic cantaloupe line with altered ethylene biosynthesis due to over expression of an anti-sense ACO. Ethylene production in ripening fruit and wounded leaves was reduced by 99 and 66%, respectively. No change in pulp ripening was observed in transgenic fruits when compared to control but the rind stayed greener due to retention of chlorophyll and had increased concentration of sucrose and citric acid (Flores et al. 2001). RNAi technique has also been used to produce tomato fruit with delayed ripening using ACO gene. The transgenic fruits developed by RNAi constructs released traces of ethylene and had a prolonged shelf life of more than 120 days (Xiong et al. 2005).

8.7.2 Transgenic Fruits with Altered Ethylene Perception

Transgenic plants have been generated with lowered ability to perceive ethylene. Ethylene receptors have been a popular starting point for these studies due to information available on the receptor genes. The mutant *etr1-1* from *Arabidopsis*, which has a point mutation in the histidine kinase domain, is responsible for ethylene insensitivity. When *etr1-1* gene was over expressed in tomato, plants showed decreased ethylene perception resulting in significant delay in fruit ripening (Wilkinson et al. 1997). Stable transformation of tomato plants with a *LeETR1* construct containing the antisense sequence for the receiver domain and 3'-untranslated portion of the gene under the control of an enhanced CaMV 35S promoter resulted in some expected and unexpected phenotypes (Whitelaw et al. 2002). In addition to reduced *LeETR1* transcript levels, the two most consistently observed phenotypes in the transgenic lines were delayed abscission and reduced plant size. Fruit coloration and softening were essentially unaffected, and all the seedlings from first generation seed displayed a normal triple response to ethylene. In another study, a wild-type form of NR, when over-expressed in tomato using constitutive 35S promoter produced plants that were less sensitive to ethylene (Ciardi et al. 2000). Conversely, antisense reduction in expression of individual receptors did not have a major effect on ethylene sensitivity. The only exception to this was antisense-reduced *LeETR4* plants, which exhibited a constitutive ethylene response and were severely affected. This is the only receptor that, when substantially reduced, causes a severe ethylene hypersensitive phenotype. The effects include epinasty, loss of flowers, and early fruit ripening without any increase in ethylene synthesis indicating that transgenic plants were more sensitive to ethylene. Hackett et al. (2000) developed transgenic *Nr* antisense plants which showed inhibition of the mutant receptor gene sufficient to allow fruit to turn red in color and to achieve wild-type levels of expression of ripening-related (*PSY1* and *ACO1*) and ethylene-responsive (*E4*) genes.

8.7.3 Transgenic Fruits with Altered Fruit Softening

The textural changes leading to softening are most prominent in climacteric fruits. These changes are brought about mostly by the solubilization and depolymerization of cell wall hemicelluloses and pectin (Fischer and Bennett 1991) by various cell wall hydrolases (Brummell and Harpster 2001). Since excessive softening is the primary cause for fruit spoilage and pathogen attack, development of transgenic fruit with reduced softening has received a lot of attention.

Tomato PG was the first cell wall hydrolase to be examined using transgenic approach. Two groups independently down-regulated PG mRNA accumulation by constitutive expression of an antisense PG transgene driven by the cauliflower mosaic virus 35S promoter (Sheehy et al. 1988; Smith et al. 1988). The studies yielded transgenic fruits, retaining only 0.5–1% of wild-type levels of PG enzyme activity (Smith et al. 1990; Kramer et al. 1992). Nevertheless, overall fruit ripening and softening was not affected. However, since the fruits had a better juice quality, they led to development and commercialization of the first transgenic fruit 'Flavr Savr' tomato marketed by Calgene Inc. in 1994. In a related study, use of the non-softening ripening-impaired *rin* mutant, in which mRNA of the endogenous PG gene accumulates at much reduced levels, was used to examine the role of PG in ripening-related cell wall metabolism (Giovannoni et al. 1989). Transgenic *rin* plants with a sense PG transgene under control of E8 promoter produced fruit with PG enzyme activity to 60% of wild type level but this did not restore softening.

Demethylation of polygalacturonans in the cell wall is accomplished by pectin methylesterase, which de-esterifies polyuronides and makes them susceptible to degradation by PG. Two groups independently suppressed PME activity in tomato by introduction of antisense PME2/PEC2 transgenes under the control of the constitutive 35S promoter. In both cases, PME2 mRNA and protein were reduced to undetectable or trace levels in fruit. The degree of pectin methyl esterification in transgenic antisense PME fruit was higher than controls by 15–40% throughout ripening (Tieman et al. 1992; Hall et al. 1993), but the fruit otherwise ripened normally (Tieman et al. 1992).

Attempts have also been made to alter the wall polymer galactoside content. In tomato, α-galactosidases are encoded by a gene family of at least seven members (*TBG1-7*) having different patterns of expression during fruit development (Smith and Gross 2000). Sense suppression by a short gene-specific region of TBG1 cDNA reduced TBG1 mRNA abundance to 10% of wild-type levels in ripe fruit, but did not reduce total exo-galactanase activity and did not affect cell wall galactose content or fruit softening (Carey et al. 2001). Antisense tomato beta-galactosidase 4 (TBG4) cDNA driven by the cauliflower mosaic virus 35S promoter resulted transgenic tomatoes with fruit firmness significantly higher than control fruit. Antisense suppression of a tomato β-galactosidase gene using TBG6 produced lines with increased fruit cracking, reduced locular space, and a doubling in the thickness of the fruit cuticle. In addition, transgenic lines exhibited a 35–39% reduction in fruit firmness at the 20 dap stage, but their texture was equivalent to the wild type at 30 dap and beyond (Moctezuma et al. 2003).

Pectate lyases catalyse the eliminative cleavage of de-esterified pectin, which is a major component of the primary cell walls of many higher plants (Carpita and Gibeaut 1993). Pectate lyase gene expression has been manipulated in transgenic strawberry fruits and suppression of the mRNA during ripening resulted in significantly firmer fruits (Jimenez-Bermudez et al. 2002). In the transgenic fruits, the highest reduction in softening occurred during the transition from the white to the red stage.

Cell wall matrix glycans undergo considerable depolymerization by cellulase (EGase) during fruit ripening which is believed to contribute substantially to fruit softening. Large, very divergent multigene families encode EGases, which in tomato consists of at least eight members. In tomato, mRNA accumulation of the highly divergent EGases *LeCel1* and *LeCel2* was suppressed individually by constitutive expression of antisense transgenes (Lashbrook et al. 1998a; Brummell et al. 1999a). In both cases, most suppressed lines showed decreased mRNA accumulation in fruit pericarp by 99% relative to wild type, without affecting the expression of the other EGase and fruit softening. In bell pepper, sense suppression of *CaCel1* reduced immunodetectable CaCell protein and extractable CMCase activity to undetectable levels in ripe fruit (Harpster et al. 2002). However, the lack of CaCell activity in suppressed fruit had no detectable effect on ripening-related matrix glycan depolymerization.

During the growth and maturation of green tomato fruit, at least six expansin genes show staggered and overlapping patterns of mRNA accumulation (Brummell et al. 1999c; Catalá et al. 2000). Suppression and overexpression of *LeExp1* mRNA resulted in altered fruit softening during ripening (Brummell et al. 1999b). Suppression of *LeExp1* protein in ripening fruit to 3% of wild-type, reduced fruit softening during ripening by 15–20%, but did not affect matrix glycan depolymerization. Interestingly, transgenic fruit showed reduced PG activity. Overexpression of *LeExp1* protein up to 3-fold of wild-type abundance did not affect polyuronide depolymerization but substantially increased fruit softening and the depolymerization of matrix glycans, including xyloglucan.

Suppression of either *LePG* or *LeExp1* expression alone results in altered softening and/or shelf life characteristics. To test whether simultaneous suppression of both *LePG* and *LeExp1* expression influences fruit texture in additive or synergistic ways, transgenic tomato lines with reduced expression of both *LePG* and *LeExp1* were crossed with each other. Fruits from the third generation of progeny, homozygous for both transgenic constructs, were analyzed for firmness and other quality traits during ripening on or off the vine. In field-grown transgenic tomato fruits, suppression of *LeExp1* or *LePG* alone did not significantly increase fruit firmness. However, fruits suppressed for both *LePG* and *LeExp1* expression were significantly firmer throughout ripening and were less susceptible to deterioration during long-term storage. Juice prepared from the transgenic tomato fruit with reduced *LePG* and *LeExp1* was more viscous than juice prepared from control fruit (Powell et al. 2003).

Apart from cell wall hydrolases, other genes have also been used to increase the post-harvest life of fruits. Deoxyhypusine synthase (DHS) mediates the first of two sequential enzymatic reactions that activate eukaryotic translation initiation factor-5A (eIF-5A) by converting a conserved Lys to the unusual amino acid, deoxyhypusine. DHS protein levels were suppressed in transgenic tomato plants by expressing the 3′-untranslated region of tomato DHS under regulation of the constitutive cauliflower mosaic virus promoter. Fruit from the transgenic plants ripened normally, but exhibited delayed post-harvest softening and senescence that correlated with suppression of DHS protein levels (Wang et al. 2005).

Lu et al. (2001) repressed the expression of a Rab11a GTPase in tomato and found that it affected several aspects of plant development. Antisense plants showed reduced ripening due to reduced levels of PG and PME indicating a role of Rab11a GTPase in trafficking of these cell wall hydrolases. Recently, Neta-Sharir et al. (2005) used transgenic analysis to show that the chloroplastic small heat-shock protein HSP21 was involved in chromoplast development. Overexpression of HSP21 in transgenic tomatoes resulted in earlier lycopene accumulation in fruits as compared to controls. Transgenics of tomato have also been prepared for understanding the role of the lipoxygenase in volatile synthesis. Antisense plants of *tomloxC*, the major lipoxygenase in fruits, showed greatly reduced volatiles of C6 type in tomato. The accumulation of these volatiles was shown to be dependent on linoleic and linolenic acids (Chen et al. 2004b).

8.7.4 Ripening-Related Promoters

As described earlier, ripening in fruit is associated with expression of hundreds of genes, many of which are specific to fruits. Some of these genes show ripening-related expression while some are strongly responsive to ethylene. Apart from understanding the *cis* acting elements that drive ethylene-responsive gene expression in fruits, the study of promoter also has a great potential for biotechnological applications where fruit specific manipulation of a gene is desired. Several fruit/ripening specific promoters from genes such as ACO (Holdsworth et al. 1987), PG (Montgomery et al. 1993), E4 (Montgomery et al. 1993), E8 (Deikman and Fischer 1988; Deikman et al. 1992), 2A11 (van Haaren and Houck 1991, 1993), TFM-7 (Santino et al. 1997) and P119 (Dunsmuir and Stott 1997) have been identified in tomato and used to produce transgenic fruits. Extensive analysis of the promoters for E4 and E8 genes has been carried out. While E8 shows fruit specificity and ethylene-responsive expression, E4 induction by ethylene is tissue independent. It was shown that at least two co-operative *cis* acting elements located between -150 and -126 and between -40 and $+65$ in E4 were necessary for ethylene responsiveness (Xu et al. 1996). In E8, the *cis* acting elements for fruit specificity and ethylene responsiveness were distinct. The region from $-1,088$ to $-2,181$ was required for ethylene responsiveness, while the fruit specific *cis* acting elements were present in the proximal portion of the promoter between -631 to -349. When this region was deleted, there was a dramatic loss of E8 gene expression in ripening fruits (Deikman et al. 1992). A DNA binding protein E4/E8BP that could bind both the E4 and E8 promoters was identified and showed expression during ripening (Coupe and Deikman 1997). The E8 promoter has been shown to be late expressing during ripening and has been used to develop transgenic tomato fruits (Sandhu et al. 2000; Lewinsohn et al. 2001; Mehta et al. 2002). The promoter region of tomato PG (TFPG) also shows significant homology to the ethylene-responsive E8 and E4 gene promoters. A 4.8 kb 5′ flanking region of TFPG greatly increased ripening specific reporter gene activity, while 1.8 kb 3′ region has a positive regulatory role in the presence of the extended 5′ region. DNA sequence of the 3.4 kb region has 400 bp imperfect reverse repeat sequences, which shows functional similarity to the regulatory regions of the ethylene regulated E4 and E8 gene promoters (Nicholass et al. 1995). The studies related to 2A11 promoter revealed that it contains at least four fruit-specific and one leaf- and fruit-active protein-binding domain (van Haaren and Houck 1991), which drive expression during fruit development and ripening. When this promoter was used to develop fruit specific RNAi mediated suppression of *det1* gene, transgenic tomato fruits accumulated enhanced amount of carotenoid and flavonoid in fruits only (Davuluri et al. 2005). In the same study, constructs containing TFM7 and P119 promoters also enhanced carotenoid and flavonoid levels specifically in the fruit tissue. This study suggests that these promoters were able to drive expression in fruit specific manner. The promoters of *LeCel2* and *LeExp1* have also been shown to contain some positively and negatively regulating regions, which can drive expression of any gene during tomato fruit ripening. The apple ACC-oxidase promoter contains elements that are located between −1,159 and −450 and direct ripening-specific gene expression in tomato fruit. The presence of a region between −450 to −1 has been shown to direct fruit but not ripening-specific gene expression (Atkinson et al. 1998). The TGTCACA motif, a novel enhancer element necessary for fruit-specific expression of the cucumisin gene reported from the melon (Yamagata et al. 2002). Though these promoters drive expression in response to ethylene, none of them contain GCC-box present in some defense-related genes that are responsive to ethylene (Ohme-Takagi and Shinshi 1990).

Though promoters from other fruits have also been isolated but in most cases analyses has been done only in the heterologous plants like tomato. The results are very encouraging but need further experimentation with respect to specificity. For example, the apple ACC-oxidase promoter contains elements located between −1159 and −450 that direct ripening-specific gene expression in tomato fruit (Atkinson et al. 1998). The region from −450 to −1 contains an element that directs fruit but not ripening-specific gene expression. Similar results were obtained for apple PG promoter showing that the region from −1 to −1460 contains positive regulatory elements that direct expression in ripe tomato fruit. The region from −2,356 to −1,460 may contain an element(s) that down-regulates expression in ripe tomato fruit, but not necessarily in apple fruit. Both the PG and ACC-oxidase promoters from apple were able to direct GUS reporter gene expression to the inner, but not outer pericarp of ripe tomato fruit. Kumar et al. (2005) have transformed banana with the 's' gene of hepatitis B surface antigen (HBsAg) under control of banana ACO promoter and shown presence of antigen in fruit tissue. Peumans et al. (2002) suggested BanTLP (thaumain-like protein) and BanGase (α-1, 3-glucanase) promoters as prospective candidates for this purpose based on the total protein analysis. Trivedi and Nath (2004) suggested the use of *MaExp1* promoter as a good candidate for fruit specific expression of genes in ripening banana fruit tissue. To investigate the transcriptional regulation of ACO genes of peach, chimeric fusions between GUS reporter gene, and Pp-ACO promoter have been constructed and introduced in tomato. When Pp-ACO promoter was used in transgenic tomato plants, same pattern of expression was observed as in peach and the GUS activity was localized in leaf blade, ovary, leaf and fruit abscission zones and pericarp, which was up-regulated by propylene and wounding. Transient expression analysis of this promoter in peach fruit suggests that it contains an ethylene-responsive element (ERE), which is responsible for the stimulation by the ethylene (Rasori et al. 2003; Moon and Callahan 2004). The ethylene-responsive promoters from tomato, apple, and peach do not share homology in their ethylene-responsive elements, which suggest that these elements might be species-specific. This observation was also made by Agius et al. (2005) with GalUR promoter from strawberry. The transient assay was used to study the activity of the tomato polygalacturonase, pepper fibrillin as well as GalUR promoter in strawberry fruits. Whereas slight activity was observed with the fibrillin promoter, no significant activity was found with the polygalacturonase promoter. The GalUR promoter in transiently transformed ripe tomato fruits showed no activity, indicating the presence of regulatory sequences specific for its function in strawberry fruit.

8.8 Conclusions

Ripening in climacteric fruits is basically a degradative process and ethylene burst during climacteric rise may be considered a death signal that prepares fruit for seed dispersal through a concerted breakdown of various cellular components. Though certain biosynthetic processes also operate during this period, these are eventually for assisting seed dispersal by acting as visual or olfactory attractants. Ethylene is perceived through a set of ethylene receptors and the signal traverses through various components in an orderly manner to reach target genes that are either expressed or repressed. A number of signaling components (molecules) have been identified and characterized in fruits as homologues to those already established in *Arabidopsis* indicating conservation in the basic signaling process. Several hundred target genes are either expressed or repressed as a consequence to ethylene signaling bringing about textural and biochemical changes in the fruit. The ripening-related changes in climacteric fruits are somewhat sudden and short lived with respect to their aesthetic and commercial value. Large-scale losses are incurred due to spoilage and short shelf life of such fruits. Attempts have been made to extend shelf life of these fruits by manipulating expression of genes involved in ethylene action. There is also greater need to identify a fruit/ripening specific promoter so that desirous genes may be specifically expressed in fruit tissue for the manipulation of ripening or value addition in fruits.

References

Abeles FB, Morgan PW, Saltveit ME (1992) Ethylene in plant biology, 2nd edn. Academic, San Diego

Adams-Phillips L, Barry C, Giovannoni J (2004) Signal transduction systems regulating fruit ripening. Trends Plant Sci 9:331–338

Agius F, Amaya I, Botella MA, Valpuesta V (2005) Functional analysis of homologous and heterologous promoters in strawberry fruits using transient expression. J Exp Bot 56:37–46

Asif MH, Nath P (2005) Expression of multiple forms of polygalacturonase gene during ripening in banana fruit. Plant Physiol Biochem 43:177–184

Atkinson RG, Bolitho KM, Wright MA, Iturriagagoitia-Bueno T, Reid SG, Ross GS (1998) Apple ACC-oxidase and polygalacturonase: ripening-specific gene expression and promoter analysis in transgenic tomato. Plant Mol Biol 38:449–460

Ayub R, Guis M, Ben Amor M, Gillot L, Roustan JP, Latche A, Bouzayen M, Pech JC (1996) Expression of ACC oxidase antisense gene inhibits ripening of cantaloupe melon fruits. Nat Biotech 14:862–865

Bahrami AR, Chen Z-H, Walker RP, Leegood RC, Gray JE (2001) Ripening-related occurrence of phosph*oenol*pyruvate carboxykinase in tomato fruit. Plant Mol Biol 47:499–506

Barry CS, Blume B, Bouzayen M, Cooper W, Hamilton AJ, Grierson D (1996) Differential expression of the 1-aminocyclopropane-1-carboxylate oxidase gene family of tomato. Plant J 19:525–535

Barry CS, Llop-Tous MI, Grierson D (2000) The regulation of 1-aminocyclopropane-1-carboxylic acid synthase gene expression during the transition from system-1 to system-2 ethylene synthesis in tomato. Plant Physiol 123:979–986

Biale JB (1964) Growth, maturation and senescence in fruits. Science 146:880–888

Bleecker AB, Kende H (2000) Ethylene: a gaseous signal molecule in plants. Annu Rev Cell Dev Biol 16:1–40

Bouquin T, Lasserre E, Paradier J, Pech JC, Balague C (1997) Wound and ethylene induction of the ACC oxidase melon gene CM-ACO1 occurs via two direct and independent transduction pathways. Plant Mol Biol 35:1029–1035

Brady CJ (1987) Fruit ripening. Annu Rev Plant Physiol Plant Mol Biol 38:155–178

Brummell DA, Harpster MH (2001) Cell wall metabolism in fruit softening and quality and its manipulation in transgenic plants. Plant Mol Biol 47:311–340

Brummell DA, Hall BD, Bennett AB (1999a) Antisense suppression of tomato endo-1,4-α-glucanase Cel2 mRNA accumulation increases the force required to break fruit abscission zones but does not affect fruit softening. Plant Mol Biol 40:615–622

Brummell DA, Harpster MH, Civello PM, Palys JM, Bennett AB, Dunsmuir P (1999b) Modification of expansin protein abundance in tomato fruit alters softening and cell wall polymer metabolism during ripening. Plant Cell 11:2203–2216

Brummell DA, Harpster MH, Dunsmuir P (1999c) Differential expression of expansin gene family members during growth and ripening of tomato fruit. Plant Mol Biol 39:161–169

Cancel JD, Larsen PB (2002) Loss-of-function mutations in the ethylene receptor ETR1 cause enhanced sensitivity and exaggerated response to ethylene in *Arabidopsis*. Plant Physiol 129:1557–1567

Carey AT, Smith DL, HalTison E, Bird CR, Gross KC, Seymour GB, Tucker GA (2001) Down-regulation of a ripening-related α-galactosidase gene (TBG1) in transgenic tomato fruits. J Exp Bot 52:663–668

Carpita NC, Gibeaut DM (1993) Structural models of primary cell walls in flowering plants: consistency of molecular structure with the physical properties of the walls during growth. Plant J 3:1–30

Catala C, Rose JKC, Bennett AB (2000) Auxin-regulated genes encoding cell wall-modifying proteins are expressed during early tomato fruit growth. Plant Physiol 122:527–534

Chang C, Bleecker AB (2004) Ethylene biology: more than a gas. Plant Physiol 136:2895–2899

Chang C, Stadler R (2001) Ethylene hormone receptor action in *Arabidopsis*. Bioassays 23:619–627

Chao QM, Rothenberg M, Solano R, Roman G, Terzaghi W, Ecker JR (1997) Activation of the ethylene gas response pathway in *Arabidopsis* by the nuclear protein ETHYLENE-INSENSITIVE and related proteins. Cell 89:1133–1144

Chen G, Alexander L, Grierson D (2004a) Constitutive expression of EIL-like transcription factor partially restores ripening in the ethylene-insensitive *Nr* tomato mutant. J Exp Bot 55:1491–1497

Chen G, Hackett R, Walker D, Taylor A, Lin Z, Grierson D (2004b) Identification of a specific isoform of tomato lipoxygenase (*TomloxC*) involved in the generation of fatty acid-derived flavor compounds. Plant Physiol 136:2641–2651

Chen YI-F, Etheridge N, Schaller GE (2005) Ethylene signal transduction. Ann Bot 95:901–915

Chervin C, El-Kereamy A, Roustan J, Latché A, Lamon J, Bouzayen M (2004) Ethylene seems required for the berry development and ripening in grape, a non-climacteric fruit. Plant Sci 167:1301–1305

Ciardi J, Klee H (2001) Regulation of ethylene-mediated responses at the level of the receptor. Ann Bot 88:813–822

Ciardi JA, Tieman DM, Lund ST, Jones JB, Stall RE, Klee HJ (2000) Response to *Xanthomonas campestris* pv. vesicatoria in tomato involves regulation of ethylene receptor gene expression. Plant Physiol 123:81–92

Clendennen SK, May GD (1997) Differential gene expression in ripening banana fruit. Plant Physiol 115:463–469

Considine MJ, Daley DO, Whelan J (2001) The expression of alternative oxidase and uncoupling protein during fruit ripening in mango. Plant Physiol 126:1619–1629

Cosgrove DJ, Gilroy S, Kao TH, Ma H, Schultz JC (2000) Plant signaling, cross talk among geneticists, physiologists, and ecologists. Plant Physiol 124:499–505

Coupe SA, Deikman J (1997) Characterization of a DNA-binding protein that interacts with 5' flanking regions of two fruit ripening genes. Plant J 11:1207–1218

Creelman RA, Mullet JE (1997) Biosynthesis and action of jasmonates in plants. Annu Rev Plant Physiol Plant Mol Biol 48:355–381

Dandekari AM, Teo G, Defilippi BG, Uratsu SL, Passey AJ, Kader AA, Stow JR, Colgan RJ, James DJ (2004) Effect of down-regulation of ethylene biosynthesis on fruit flavor complex in apple fruit. Transgenic Res 13:373–384

Davies C, Boss PK and Robinson SP (1997) Treatment of grape berries, a non climacteric fruit with a synthetic auxin, retards ripening and alters the expression of developmentally regulated genes. Plant Physiol 115:1155–1161

Davuluri GR, van Tuinen A, Fraser PD, Manfredonia A, Newman R, Burgess D, Brummell DA, King SR, Palys J, Uhlig J, Bramley PM, Pennings HM, Bowler C (2005) Fruit-specific RNAi-mediated suppression of DET1 enhances carotenoid and flavonoid content in tomatoes. Nat Biotech 23:890–895

Defilippi BG, Kader AA, Dandekar AM (2005) Apple aroma: alcohol acyltransferase, a rate limiting step for ester biosynthesis, is regulated by ethylene. Plant Sci 168:1199–1210

Deikman J, Fischer RL (1988) Interaction of a DNA binding factor with the 5'-flanking region of an ethylene-responsive fruit-ripening gene from tomato. EMBO J 7:3315–3320

Deikman J, Kline R, Fische RL (1992) Organization of ripening and ethylene regulatory regions in a fruit-specific promoter from tomato (*Lycopersicon esculentuns*). Plant Physiol 100:2013–2017

DellaPenna D, Lincoln JE, Fischer RL, Bennett AB (1989) Transcriptional analysis of polygalacturonase and other ripening-associated genes in Rutgers, *rin*, *nor*, and *Nr* tomato fruit. Plant Physiol 90:1372–1377

Ding C, Wang CY (2003) The dual effects of methyl salicylate on ripening and expression of ethylene biosynthesis genes in tomato fruit. Plant Sci 164:589–596

Dong JG, Fernandezmaculet JC, Yang SF (1992) Purification and characterization of 1-aminocyclopropane-1-carboxylate oxidase from apple fruit. Pro Natl Acad Sci USA 89:9789–9793

Dunsmuir P, Stott J (1997) P119 promoter and their uses. US Patent 5,633,440

El-Sharkawy I, Jones B, Li ZG, Lelievre JM, Pech JC, Latche A (2003) Isolation and characterization of four ethylene perception elements and their expression during ripening in pears (*Pyrus communis* L) with/without cold requirement. J Exp Bot 54:1615–1625

Fan X, Matthies JP, Fellman JK (1998) A role for jasmonate in climacteric fruit ripening. Planta 204:444–449

Fischer RL, Bennett AB (1991) Role of cell wall hydrolases in fruit ripening. Annu Rev Plant Physiol Plant Mol Biol 42:675–703

Flores F, Martinez-Madrid MC, Sanchez-Hidalgo FJ, Romojaro F (2001) Differential rind and pulp ripening of transgenic antisense ACC oxidase melon. Plant Physiol Biochem 39:37–43

Flores F, El Yahaoui F, de Billerbeck G, Romojaro F, Latche A, Bouzayen M, Pech JC, Ambid C (2002) Role of ethylene in the biosynthetic pathway of aliphatic ester aroma volatiles in *Charentais Cantaloupe* melons. J Exp Bot 53:201–206

Fung RWM, Langenkaemper G, Gardner RC, MacRae E (2003) Differential expression within an SPS gene family. Plant Sci 164:459–470

Gilad A, Amitai-Zeigerson H, Scolnik PA, Bar-Zvi D (1997) *Asr1*, a tomato water stressed regulated gene: genomic organization, developmental regulation and DNA binding activity. Acta Hort 447:447–454

Giovannoni JJ (2001) Molecular biology of fruit maturation and ripening. Annu Rev Plant Physiol Plant Mol Biol 52:725–749

Giovannoni JJ (2004) Genetic regulation of fruit development and ripening. Plant Cell 16:s170–s180

Giovannoni JJ, Della Penna D, Bennett AB, Fischer RL (1989) Expression of a chimeric polygalacturonase gene in transgenic *rin* (ripening inhibitor) tomato fruit results in polyuronide degradation but not fruit softening. Plant Cell 1:53–63

Good X, Kellogg JA, Wagoner W, Langoff D, Matsumura W, Bestwick RK (1994) Reduced ethylene synthesis by transgenic tomatoes expressing S-adenosylmethionine hydrolase. Plant Mol Biol 26:781–790

Goren R, Dagan E, Sagee O, Riov J, Yang SF (1993) Abscission in citrus leaf explants: role of ABA induced ethylene. Acta Hort 329:43–50

Gray J, Picton S, Shabeer J, Schuch W, Grierson D (1992) Molecular biology of fruit ripening and its manipulation with antisense genes. Plant Mol Biol 19:69–87

Gray WM (2004) Hormonal regulation of plant growth and development. PloS Biology 1.2. 9:1270–1273

Grierson D (1987) Senescence in fruits. Hort Sci 22:859–862

Griffiths A, Barry C, Alpuche-Solis AG, Grierson D (1999) Ethylene and developmental signals regulate expression of lipoxygenase genes during tomato fruit ripening. J Exp Bot 50:793–798

Guo H, Ecker JR (2003) Plant responses to ethylene gas are mediated by $SCF^{EBF1/EBF2}$-dependent proteolysis of EIN3 transcription factor. Cell 115:667–677

Guo H, Ecker JR (2004) The ethylene signaling pathway: new insights. Curr Opin Plant Biol 7:40–49

Hackett RM, Ho C, Lin Z, Foote HCC, Fray RG, Grierson D (2000) Antisense inhibition of the Nr gene restores normal ripening to the tomato Never-ripe mutant, consistent with the ethylene receptor inhibition model. Plant Physiol 124:1079–1085

Hadfield KA, Dang T, Guis M, Pech JC, Bouzayen M, Bennett AB (2000) Characterization of ripening-regulated cDNAs and their expression in ethylene-suppressed *Charentais* melon fruit. Plant Physiol 122:977–983

Hall LN, Tucker GA, Smith CS, Watson CP, Seymour GB, Bundick Y, Boniwell JM, Fletcher JD, Ray JA, Schuch W, Bird C, Glierson D (1993) Antisense inhibition of pectin esterase gene expression in transgenic tomatoes. Plant J 3:121–129

Hamilton AJ, Lycett GW, Grierson D (1990) Antisense gene that inhibits synthesis of the hormone ethylene in transgenic plants. Nature 346:284–287

Harpster MH, Dawson DM, Nevins DJ, Dunsmuir P, Brummell DA (2002) Constitutive over expression of a ripening-related pepper endo-1,4-beta-glucanase in transgenic tomato fruit does not increase xyloglucan depolymerization or fruit softening. Plant Mol Biol 50:357–369

Harrison EP, McQueen-Mason SJ, Manning K (2001) Expression of six expansion genes in relation to extension activity in developing strawberry fruit. J Exp Bot 52:1437–1446

Hiwasa K, Rose JKC, Nakano R, Inaba A, Kubo Y (2003) Differential expression of seven á-expansin genes during growth and ripening of pear fruit. Physiol Plant 117:564–572

Holdsworth MJ, Schuch W, Grierson D (1987) Nucleotide sequence of an ethylene-related gene from tomato. Nucleic Acids Res 15:10600

Hong S, Kim I, Yang DC, Chung W (2002) Characterization of an abscisic acid responsive gene homologue from *Cucumis melo*. J Exp Bot 378:2271–2272

Hua J, Meyerowitz EM (1998) Ethylene responses are negatively regulated by a receptor gene family in *Arabidopsis thaliana*. Cell 94:261–271

Irifune K, Nishida T, Egawa H, Nagatani A (2004) Pectin methylesterase inhibitor cDNA from kiwi fruit. Plant Cell Rep 23:333–338

Ishiki Y, Oda A, Yaegashi Y, Orihara Y, Arai T, Hirabayashi T, Nakagawa H, Sato T (2000) Cloning of an auxin-responsive 1-aminocyclopropane-1-carboxylate synthase gene (*Cme-ACS2*) from melon and expression of ACS genes in etiolated melon seedlings and melon fruits. Plant Sci 159:173–181

Jiang Y, Joyce DC, Macnish AJ (2000) Effect of abscisic acid on banana fruit ripening in relation to the role of ethylene. J Plant Growth Regul 19:106–111

Jimenez-Bermudez S, Redondo-Nevado J, Munoz-Blanco J, Caballero JL, Lopez-Aranda JM, Valpuesta V, Pliego-Alfaro F, Quesada MA, Mercado JA (2002) Manipulation of strawberry fruit softening by antisense expression of a pectate lyase gene. Plant Physiol 128:751–759

Kalifa Y, Gilad A, Konrad Z, Zaccai M, Scolnik PA, Bar-Zvi D (2004) The water- and salt-stress-regulated *Asr1* (abscisic acid stress ripening) gene encodes a zinc-dependent DNA binding protein. Biochem J 381:373–378

Kende H (1993) Ethylene biosynthesis. Annu Rev Plant Physiol Plant Mol Biol 44:283–307

Klee HJ (2004) Ethylene signal transduction. Moving beyond *Arabidopsis*. Plant Physiol 135:660–667

Klee HJ, Hayford MB, Kretzmer KA, Barry GF, Kishmore GM (1991) Control of ethylene synthesis by expression of a bacterial enzyme in transgenic tomato plants. Plant Cell 3:1187–1193

Kramer M, Sanders R, Bolkan H, Waters C, Sheehy RE, Hiatt WR (1992) Postharvest evaluation of transgenic tomatoes with reduced levels of polygalacturonase: processing, firmness and disease resistance. Postharvest Biol Technol 1:241–255

Kumar GB, Ganapathi TR, Revathi CJ, Srinivas L, Bapat VA (2005) Expression of hepatitis B surface antigen in transgenic banana plants. Planta 2005 222:484–493

Lashbrook CC, Giovannoni J, Hall BD, Fischer RL, Bennett AB (1998a) Transgenic analysis of tomato endo-α-1,4glucanase gene function. Role of *cell* in floral abscission. Plant J 13:303–310

Lashbrook CC, Tieman DM, Klee HJ (1998b) Differential regulation of the tomato ETR gene family throughout plant development. Plant J 15:243–252

Leclercq J, Adams-Philips L, Zegzouti H, Jones B, Latche A, Giovannoni J, Pech JC, Bouzayen M (2002) LeCTR1, a tomato CTR1-like gene, demonstrates ethylene signaling ability in *Arabidopsis* and novel expression patterns in tomato. Plant Physiol 130:1132–1142

Leclercq J, Ranty B, Sanchez-Ballesta MT, Li Z, Jauneau A, Pech JC, Lathche A, Ranjeva R, Bouzayen M (2005) Molecular and biochemical characterization of *LeCRK1*, a ripening-associated tomato CDPK-related kinase. J Exp Bot 56:25–35

Lelievre JM, Latche A, Jones B, Bouzayen M, Pech JC (1997) Ethylene and fruit ripening. Physiol Plant 101:727–739

Leon P, Sheen J (2003) Sugar and hormone connection. Trends Plant Sci 8:110–116

Leslie CA, Romani RG (1988) Inhibition of ethylene biosynthesis by salicylic acid. Plant Physiol 88:833–837

Lewinsohn E, Schalechet F, Wilkinson J, Matsui K, Tadmor Y, Nam KH, Amar O, Lastochkin E, Larkov O, Ravid U, Hiatt W, Gepstein S, Pichersky E (2001) Enhanced levels of the aroma and flavor compound S-linalool by metabolic engineering of the terpenoid pathway in tomato fruits. Plant Physiol 127:1256–1265

Li N, Parsons BL, Liu D, Mattoo AK (1992) Accumulation of wound inducible ACC synthase transcript in tomato fruit is inhibited by salicylic acid and polyamines. Plant Mol Biol 18:477–487

Liu C, Tian Y, Shen Q, Jiang H, Ju R, Yan T, Liu C, Mang K (1998) Cloning of 1-aminocyclo-propane-1-carboxylate (ACC) synthetase cDNA and the inhibition of fruit ripening by its antisense RNA in transgenic tomato plants. Chin J Biotech 14:75–84

Liu K, Kang B, Jiang H Moore SL, Li H, Watkins CB, Setter TL, Jahn MA (2005) GH3 like gene, *CcGh3*, isolated from *Capsicum chinense* L. fruit is regulated by auxin and ethylene. Plant Mol Biol 58:447–464

Liu Y, Schiff M, Dinesh-Kumar SP (2002) Virus induced gene silencing in tomato. Plant J 31:777–786

Lohani S, Trivedi PK, Nath P (2004) Changes in activities of cell wall hydrolases during ethylene induced ripening in banana: effect of 1-MCP, ABA and auxin. Postharvest Biol Technol 31:119–126

Lorenzo O, Piqueras R, Sanchez-Serrano JJ, Solano R (2003) ETHYLENE RESPONSE FACTOR1 integrates signals from ethylene and jasmonate pathways in plant defence. Plant Cell 15:165–178

Lu CG, Zainal Z, Tucker GA, Lycett GW (2001) Developmental abnormalities and reduced fruit softening in tomato plants expressing an antisense Rab11 GTPase gene. Plant Cell 13:1819–1833

Lui Y, Hoffman NE, Yang SF (1985) Promotion by ethylene of the capacity to convert 1-aminocyclopropane-1-carboxylic acid to ethylene in preclimacteric tomato and cantaloupe fruit. Plant Physiol 77:407–411

Mattoo AK, Suttle CS (1991) The plant hormone ethylene. CRC Press, Boca Raton, FL

Mbeguie A, Mbeguie D, Gomez RM, Fils-Lucaon B (1997) Molecular cloning and nucleotide sequence of an abscisic acid-stress-ripening induced (ASR)-like protein from apricot fruit (acc. No U93164). Gene expression during ripening (PGR97–166). Plant Physiol 115:1288

McMurchie EJ, McGlasson WB, Eaks IL (1972) Treatment of fruit with propylene gives information about the biogenesis of ethylene. Nature 237:235–236

Mehta RA, Cassol T, Li N, Ali N, Handa AK, Mattoo AK (2002) Engineered polyamine accumulation in tomato enhances phytonutrient content, juice quality, and vine life. Nat Biotech 20:613–618

Mita S, Kawamura S, Yamawaki K, Nakamura K, Hyodo H (1998) Differential expression of genes involved in the biosynthesis and per-ception of ethylene during ripening of passion fruit (*Passiflora edulis* Sims). Plant Cell Physiol 39:1209–1217

Moctezuma E, Smith DL, Gross KC (2003) Antisense suppression of a beta-galactosidase gene (TBG6) in tomato increases fruit cracking. J Exp Bot 54:2025–2033

Montgomery J, Pollard V, Deikman J, Fischer RL (1993) Positive and negative regulatory regions control the spatial distribution of polygalacturonase transcription in tomato fruit pericarp. Plant Cell 5:1049–1062

Moon H, Callahan AM (2004) Developmental regulation of peach ACC oxidase promoter – GUS fusions in transgenic tomato fruits. J Exp Bot 55:1519–1528

Nakatsuka A, Murachi S, Okunishi H, Shiomi S, Nakano R, Kubo Y, Inaba A (1998) Differential expression and internal feedback regulation of 1-aminocyclopropane-1-carboxylate synthase, 1-aminocyclopropane-1-carboxylate oxidase, and ethylene receptor genes in tomato fruit during development and ripening. Plant Physiol 118:1295–1305

Neta-Sharir I, Isaacson T, Lurie S, Weiss D (2005) Dual role for tomato heat shock protein 21: protecting photosystem II from oxidative stress and promoting color changes during fruit maturation. Plant Cell 17:1829–1838

Nicholass FJ, Smith CJS, Schuch W, Bird CR, Grierson D (1995) High levels of ripening-specific reporter gene expression directed by tomato fruit polygalacturonase gene-flanking regions. Plant Mol Biol 28:423–435

Oeller PW, Lu MW, Taylor LP, Pike DA, Theologis A (1991) Reversible inhibition of tomato fruit senescence by antisense RNA. Science 254:437–439

Oetiker JH, Olson DC, Shiu OY, Yang SF (1997) Differential induction of seven 1-aminocyclopropane-1-carboxylate synthase genes by elicitor in suspension cultures of tomato (*Lycopersicon esculentum*). Plant Mol Biol 34:275–286

Ohme-Takagi M, Shinshi H (1990) Structure and expression of tobacco α-1,3-glucanase gene. Plant Mol Biol 15:941–946

Ouaked F, Rozhon W, Lecourieux D, Hirt H (2003) A MAPK pathway mediates ethylene signalling in plants. EMBO J 22:1282–1288

Parikh HR, Palejwala VA, Modi VV (1986) Effect of abscisic acid on ripening of mangoes. Indian J Exp Biol 24:722–725

Pathak N, Asif MH, Dhawan P, Srivastava MK, Nath P (2003) Expression and activities of ethylene biosynthesis enzymes during ripening in banana fruits and effect of 1-MCP treatment. Plant Growth Regul 40:11–19

Payton S, Fray RG, Brown S, Grierson D (1996) Ethylene receptor expression is regulated during fruit ripening, flower senescence and abscission. Plant Mol Biol 31:1227–1231

Perez AG, Sanz C, Richardson DG, Olias JM (1993) Methyle jasmonate vapours promote α-carotene synthesis and chlorophyll degradation in Golden delicious apple peel. J Plant Growth Regul 12:163–167

Peumans WJ, Proost P, Swennen RL, van Damme EJ (2002) The abundant class III chitinase homolog in young developing banana fruits behaves as a transient vegetative storage protein and most probably serves as an important supply of amino acids for the synthesis of ripening-associated proteins. Plant Physiol 130:1063–1072

Polder G, van der Heijden GWAM, van der Voet H, Young IT (2004) Measuring surface distribution of carotenes and chlorophyll in ripening tomatoes using imaging spectrometry. Postharvest Biol Technol 34:117–129

Powell AL, Kalamaki MS, Kurien PA, Gurrieri S, Bennett AB (2003) Simultaneous transgenic suppression of LePG and LeExp1 influences fruit texture and juice viscosity in a fresh market tomato variety. J Agric Food Chem 51:7450–7455

Pua EC, Chandramouli S, Han P, Liu P (2003) Malate synthase gene expression during fruit ripening of Cavendish banana (*Musa acuminata* cv. Williams). J Exp Bot 54:309–316

Rasori A, Ruperti B, Bonghi C, Tonutti P, Ramina A (2002) Characterization of two putative ethylene receptor genes expressed during peach fruit development and abscission. J Exp Bot 53:2333–2339

Rasori A, Bertolasi B, Furini A, Bonghi C, Tonutti P, Ramina A (2003) Functional analysis of peach ACC oxidase promoters in transgenic tomato and in ripening peach fruit. Plant Sci 165:523–530

Rose JKC, Lee HH, Bennett AB (1997) Expression of a divergent expansin gene is fruit-specific and ripening-regulated. Proc Natl Acad Sci USA 94:5955–5960

Ross G, Knighton ML, Lay-Yee M (1992) An ethylene-related cDNA from ripening apples. Plant Mol Biol 19:231–238

Sandhu JS, Krasnyanski SF, Domier LL, Korban S, Osadjan MD, Buetow DE (2000) Oral immunization of mice with transgenic tomato fruit expressing respiratory syncytial virus-F protein induces a systemic immune response. Transgenic Res 9:127–135

Sane VA, Chourasia A, Nath P (2005) Softening in mango (*Mangifera indica* var Dashehari) is correlated with the expression of the early ethylene-responsive, ripening-related expansin gene, MiExpA1. Postharvest Biol Technol (in press)

Saniewski M, Czapski J (1983) The effect of methyl jasmonate on lycopene and α-carotene accumulation in ripening red tomatoes. Experentia 39:1373–1374

Saniewski M, Czapski J, Nowacki J, Lange E (1987) The effect of methyl jasmonate on ethylene and 1-aminocyclopropane-1-carboxylic acid production in apple fruits. Bio Plant 29:199–203

Santino CG, Stanford GL, Conner TW (1997) Developmental and transgenic analysis of two tomato fruit enhanced genes. Plant Mol Biol 33:405–416

Sato-Nara K, Yuhashi K, Higashi K, Hosoya K, Kubota M, Ezura H (1999) Stage- and tissue-specific expression of ethylene receptor homologue genes during fruit development in muskmelon. Plant Physiol 119:321–329

Schaller GE, Kieber JJ (2002) Ethylene. In: Somerville C, Meyerowitz E (eds) The *Arabidopsis* book. vol DOI/10.1199/tab.0071. Am Soc Plant Biol, Rockville, MD

Seymour GB, Taylor JE, Tucker GA (1993) Biochemistry of fruit ripening. Chapman and Hall, London

Sheehy RE, Kramer M, Hiatt WR (1988) Reduction of polygalacturonase activity in tomato fruit by antisense RNA. Proc Natl Acad Sci USA 85:8805–8809

Shiomi S, Yamamoto M, Ono T, kakiuchi K, Nakamoto J, Nakatsuka A, Kubo Y, Nakamura R, Inaba A, Imaseki H (1998) cDNA cloning of ACC synthase and ACC oxidase in cucumber fruit and their differential expression by wounding and auxin. J Jpn Soc Hort Sci 67:685–692

Sinha A, Asif MH, Nath P, Pathre UV (2002) Ethylene and sugar signaling in plants: do signal transduction cascades intersect each other ay any point? In: Nath P, Mattoo AK, Ranade SA, Weil JH (eds) Molecular insight in plant biology. Oxford and IBH Publ, New Delhi and Science Publ UK and USA, pp 169–188

Slovin JP, Cohen JD (1993) Auxin metabolism in relation to fruit ripening. Acta Hort 329:84–89

Smith CJS, Watson CP, Ray J, Bird CR, Morris PC, Schuch W, Grierson D (1988) Antisense RNA inhibition of polygalacturonase gene expression in transgenic tomatoes. Nature 334:724–726

Smith CJS, Watson CP, Morris PC, Bird CR, Seymour GB, Gray JE, Arnold C, Tucker GA, Schuch W, Harding S, Grierson D (1990) Inheritance and effect on ripening of antisense polygalacturonase genes in transgenic tomatoes. Plant Mol Biol 14:369–379

Smith DL, Gross KC (2000) A family of at least seven α-galactosidase genes is expressed during tomato fruit development. Plant Physiol 123:1173–1183

Solano R, Stepanova A, Chao QM, Ecker JR (1998) Nuclear events in ethylene signaling: a transcriptional cascade mediated by ETHYLENE-INSENSITIVE3 and ETHYLENE-RESPONSE-FACTOR1. Gene Dev 12:3703–3714

Spanu P, Grosskopf DG, Felix G, Boller T (1994) The apparent turnover of 1-aminocyclopropane-1-carboxylate synthase in tomato cells is regulated by protein phosphorylation and dephosphorylation. Plant Physiol 106:529–535

Tatsuki M, Mori H (2001) Phosphorylation of tomato 1-aminocyclopropane-1-carboxylic acid synthase, LE-ACS2, at the C-terminal region. J Biol Chem 276:28051–28057

Tesniere C, Pradal M, El-Kereamy A, Torregrosa L, Chatelet P, Roustan JP, Chervin C (2004) Involvement of ethylene signalling in a non-climacteric fruit: new elements regarding the regulation of ADH expression in grapevine. J Exp Bot 55:2235–2240

Tieman DM, Harriiman RW, Ramamohan G, Handa AK (1992) An antisense pectin methylesterase gene alters pectin chemistry and soluble solids in tomato fruit. Plant Cell 4:667–679

Tieman DM, Ciardi JA, Taylor MG, Klee HJ (2001) Members of the tomato LeEIL (EIN3-like) gene family are functionally redundant and regulate ethylene responses throughout plant development. Plant J 26:47–58

Tieman DV, Taylor MG, Ciardi JA, Klee HJ (2000) The tomato ethylene receptors NR and LeETR4 are negative regulators of ethylene response and exhibit functional compensation within a multigene family. Proc Natl Acad Sci USA 97:5663–5668

Tigchelaar EC, McGlasson WB, Buescher RW (1978) Genetic regulation of tomato ripening. Hort Sci 13:508–513

Tournier B, Sanchez-Ballesta MT, Jones B, Pesquet E, Regad F, Latche A, Pech JC, Bouzayen M (2003) New members of the tomato ERF family show specific expression pattern and diverse DNA binding capacity to the GCC box element. FEBS Lett 550:149–154

Trainotti L, Pavanello A, Casadoro G (2005) Different ethylene receptors show an increased expression during the ripening of strawberries: does such an increment imply a role for ethylene in the ripening of these non-climacteric fruits? J Exp Bot 56:2037–2046

Trivedi PK, Nath P (2004) MaExp1, an ethylene-induced expansin from ripening banana fruit. Plant Sci 167:1351–1358

Van Haaren MJ, Houck CM (1991) Strong negative and positive regulatory elements contribute to the high-level fruit-specific expression of the tomato 2A11 gene. Plant Mol Biol 17:615–630

Van Haaren MJ, Houck CM (1993) A functional map of the fruit-specific promoter of the tomato 2A11 gene. Plant Mol Biol 21:625–640

Von Loesecke HW (1950) Bananas: chemistry, physiology, technology. Interscience, New York, pp 108–109

Vrebalov J, Ruezinsky D, Padmanabhan V, White R, Medrano D, Drake R, Schuch W, Giovannoni J (2002) A MADS-box gene necessary for fruit ripening at tomato Ripening-inhibitor (Rin) locus. Science 296:343–346

Wang TW, Zhang CG, Wu W, Nowack LM, Madey E, Thompson JE (2005) Antisense suppression of deoxyhypusine synthase in tomato delays fruit softening and alters growth and development. Plant Physiol 138:1372–1382

Watada NL, Herner RC, Kader AA, Romani RJ, Staby GL (1984) Terminology for the description of developmental stages of horticultural crops. Hort Sci 19:20–21

Whitelaw CA, Lyssenko NN, Chen L, Zhou D, Mattoo AK, Tucker ML (2002) Delayed abscission and shorter internodes correlate with a reduction in the ethylene receptor LeETR1 transcript in transgenic tomato. Plant Physiol 128:978–987

Whittaker DJ, Smith GS, Gardner RC (1997) Expression of ethylene biosynthetic genes in *Actinidia chinensis* fruit. Plant Mol Biol 34:45–55

Wilkinson J, Lanahan M, Yen H, Giovannoni J, Klee H (1995)An ethylene-inducible component of signal transduction encoded by Never-ripe. Science 270:1807–1809

Wilkinson JQ, Lanahan MB, Clark DG, Bleeker AB, Chang C, Meyerowitz EM, Klee HJ (1997) A dominant mutant receptor for *Arabidopsis* confers ethylene insensitivity in heterologous plants. Nature Biotech 15:444–447

Wong WS, Ning W, Xu PL, Kung SD, Yang SF, Li N (1999) Identification of two chilling-regulated 1-aminocyclopropane-1-carboxylate synthase genes from citrus (*Citrus sinensis* Osbeck) fruit. Plant Mol Biol 41:587–600

Xiong AS, Yao QH, Peng RH, Li X, Han PL, Fan HQ (2005) Different effects on ACC oxidase gene silencing triggered by RNA interference in transgenic tomato. Plant Cell Rep 23:639–646

Xu R, Goldman S, Coupe S, Deikman J (1996) Ethylene control of E4 transcription during tomato fruit ripening involves two cooperative cis-elements. Plant Mol Biol 31:1117–1127

Yamagata H, Yonesu K, Hirata A, Aizono Y (2002) TGTCACA motif is a novel cis-regulatory enhancer element involved in fruit-specific expression of the cucumisin gene. J Biol Chem 29:11582–11590

Yamamoto M, Miki T, Ishiki Y, Fujinami K, Yanagisawa Y, Nakagawa H, Ogura N, Hirabayashi T, Sato T (1995) The synthesis of ethylene in melon fruit during the early-stage of ripening. Plant Cell Physiol 36:591–596

Yanagisawa S, Yoo SD, Sheen J (2003) Differential regulation of EIN3 stability by glucose and ethylene signaling in plants. Nature 425:521–525

Yang SF, Hoffman NE (1984) Ethylene biosynthesis and its regulation in higher plants. Annu Rev Plant Physiol 35:155–189

Zegzouti H, Jones B, Frasse P, Marty C, Maitre B, Latche A, Pech J-C, Bouzayen M (1999) Ethylene-regulated gene expression in tomato fruit: characterization of novel ethylene-responsive and ripening-related genes isolated by differential display. Plant J 18:589–600

Zhou DB, Kalaitzis P, Mattoo AK, Tucker ML (1996) The mRNA for an ETR1 homologue in tomato is constitutively expressed in vegetative and reproductive tissues. Plant Mol Biol 30:1331–1338

9 Ethylene Involvement in Photosynthesis and Growth

NAFEES.A. KHAN

9.1 Introduction

To distinguish plant hormones from that of animal, the term phytohormone has been used. Phytohormones are extremely important agents required in the control of developmental activities. They regulate expression of the intrinsic genetic potential of plants. Although the mechanism is not well understood, the control of genetic expression has been demonstrated for the phytohormones at the transcription and translation level (Arteca 1997). Also, hormones receptors binding proteins have been identified on membrane surface that are specific for some phytohormones. The type and abundance of these proteins appear to be important in determining the sensitivity of the tissue to phytohormones.

Of the five groups of hormones, each affects photosynthesis by modifying plant growth or enzymes of photosynthesis or by the interaction with other plant hormones. It is known that a particular process is affected by more than one group of hormones. Among the phytohormones, the interactions of gibberellins, abscisic acid, and ethylene are considered to set leaf potential for photosynthesis (Fig. 9.1).

9.2 Plant Growth Regulators and Photosynthetic Responses

Plant hormones are important in delivering photosynthetic responses. Phytohormones play an essential role in integrating many aspects of plant development and response to the environment. Regulation of hormonally controlled events can occur at the level of biosynthesis, catabolism, or perception (Trewavas 1983; Bradford and Trewavas 1994). Phytohormones improve the physiological efficiency of plants by modifying the balance between photosynthesis and respiration (Arteca and Dong 1981; Zerbe and Wild 1981; Makeev et al. 1992). Increased rates of photosynthesis per unit leaf area have been observed after the application of phytohormones on different

---

Department of Botany, Aligarh Muslim University, Aligarh 202002, India

Ethylene Action in Plants
(ed. by N.A. Khan)
© Springer-Verlag Berlin Heidelberg 2006

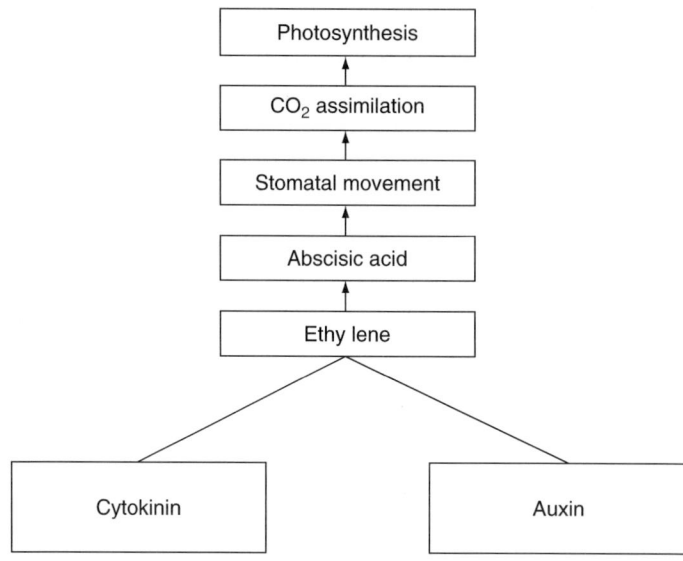

Fig. 9.1. Interrelation of phytohormones with photosynthesis

species (Child et al. 1985; Liu et al. 1993; Yang et al. 1994; Khan 1996b; Khan et al. 1996, 2000). Phytohormones can affect photosynthetic CO_2 uptake either by affecting stomatal aperture or by affecting the activity of photosynthetic enzymes (Foroutan-Pour et al. 1997).

Varied information on the effects of gibberellic acid (GA) on photosynthesis has been reported in the literature. Some reports advocate that GA enhances photosynthesis (Khan 1996a; Khan et al. 1996, 2002b; Yuan and Xu 2001) but others report decreased rate of photosynthesis (Dijkstra et al. 1990). The use of GA biosynthesis inhibitors has been found to stimulate photosynthesis (Sairam et al. 1991; Thetford et al. 1995) or reduce the photosynthetic rate (Bode and Wild 1984; Heide et al. 1985). The studies employing GA mutants showed no difference in the photosynthetic rate between wild-type tomato plants and GA mutants. These contradictory reports may be due to several reasons, including measurement and expression of photosynthesis, and the application of GA or GA inhibitors (Nagel and Lambers 2002). The other mechanism presumed to be operative in GA-mediated photosynthetic responses is the involvement of other hormones, as perception of one plant hormone affects the concentrations, perception, and responses to others (Khan et al. 2003). In this aspect, the interaction of ethylene, GA, and abscisic acid are possible because ethylene has a role in photosynthesis through its effect on leaf area index and light interception (Subrahmanyam and Rathore 1992; Khan et al. 2000). Another reason to link ethylene's role in photosynthesis is the similarity of formation of carbamate from CO_2. As in ribulose-1,5 bisphosphate carboxylase (RuBPC), CO_2 binds a Lys residue (Andrews and Lorimer 1987) forming a carbamate (Lorimer and Miziorka 1980), similar carbamate formation (Fernandez-Maculet et al. 1993) and a single Lys (Ververides and Dilley 1994) involvement of 1-aminocyclopropane carboxylic acid (ACC) oxidase (an enzyme responsible for ethylene release) activation by CO_2 has been suggested. Besides this, ethylene-induced change in abscisic acid biosynthesis may function in causing alteration in photosynthesis through the established role of abscisic acid in stomatal movement and gas exchange. Study on deepwater rice has shown that ethylene accumulation in the submerged tissue led to a decrease in abscisic acid levels (Kende and Zeevaart 1997; Grossmann and Huesen 2001) and an increase in the responsiveness of the tissue to gibberellic acid. Other reports on the interaction of ethylene and abscisic acid also show that applied abscisic acid inhibits ethylene evolution (Tan and Thimann 1989). Abscisic acid antagonist treatment increases the rate of ethylene evolution from maize seedlings (Spollen et al. 2000) and abscisic acid-deficient *Arabidopsis* mutant has been found to exhibit an increase in ethylene evolution (Rakitina et al. 1994). The hormone auxin also influences ethylene evolution. Increased auxin supply stimulates ethylene evolution through de novo synthesis of ACC synthase, the first enzyme in ethylene biosynthesis. Similarly, cytokinins at a high concentration increase ethylene evolution. It is not known whether endogenous abscisic acid functions to restrict synthesis of ethylene precursor ACC or the conversion of ACC to ethylene. This could be resolved by examining the effects of abscisic acid deficiency in ethylene-deficient or ethylene-insensitive plants (Sharp and Le Noble 2002).

9.3 Ethylene in Photosynthesis and Growth of Mustard (*Brassica juncea*)

Ethylene is a plant hormone that influences many aspects of plant growth and development (Mattoo and White 1991; Abeles et al. 1992; Wang et al. 2002). Its biosynthetic pathway has been well established (Kende 1993). In higher plants, ethylene is synthesized from methionine through S-adenosylmethionine. ACC synthase is the enzyme involved in catalyzing the first step of ethylene biosynthesis. It catalyses the conversion of S-adenosylmethionine to ACC, which in turn is converted to ethylene by the mediation of ACC oxidase. Both the enzymes ACC synthase and ACC oxidase are important in ethylene biosynthesis.

Since the discovery of ethylene, its function has been implicated to fruit ripening and, therefore, generally termed as the 'ripening hormone'. In recent years, considerable research has been conducted on understanding the biochemical and physiological bases of the effects of ethylene in higher plants. Several reports indicate that ethylene is involved in every aspect of plant growth and development and CO_2 assimilation. The effects of ethylene are diverse, depending on its concentration in the tissue (Hussain et al. 1999).

Despite its simple two-carbon structure, ethylene is a potential modulator of growth and development (Abeles et al. 1992; Kieber et al. 1993; Ecker 1995; Smalle and van der Straeten 1997). It is a simple, readily diffusible hormone with an important role in integrating developmental signals and responses to external stimuli (Ciardi and Klee 2001). The production of ethylene is tightly regulated by internal signals during development and in response to external environmental stimuli. In 1901, the Russian botanist Neljubow was the first to recognize the growth regulatory properties of ethylene. Later, in 1930, ethylene was recognized to have a vide variety of effects on plants and in 1934, Gane in England first obtained positive proof that ethylene was a natural plant product.

Ethylene is the simplest olefin, which exists in the gaseous state under normal physiological conditions. Ethylene-generating commercial chemical is ethephon or ethrel (2-chloroethyl phosphonic acid, $ClCH_2CH_2PO_3H_2$) (Warner and Leopold 1969). Ethephon is the most important and versatile ethylene-releasing agent marketed and registered for more than 20 crops. It is a synthetic plant growth regulator that undergoes chemical biodegradation in cell cytoplasm at pH greater than 4.1 to release ethylene (Urwiller and Stutte 1986; Kasele et al. 1995). Not all plants respond to ethylene in the same manner nor are they equally sensitive to this phytohormone. For example, a higher concentration of ethylene is required to inhibit stem elongation in monocotyledons than dicotyledons (Abeles et al. 1992), while exposure to ethylene stimulates the growth of other plants, like rice (Raskin and Kende 1984). Similarly, the response of tissue to ethylene is dictated by the concentration to which they are exposed. In the control of growth, ethylene is found to promote the reorientation of cortical microtubules, thereby possibly controlling elongation (Shibaoka 1994). Moreover, a role of ethylene has been suggested in regulating the expression of cell wall peroxidase involved in the control of wall extensibility and cell growth (Ridge and Osborne 1971). Regulating the level of peroxidase activity by suitable concentration of ethylene influenced the direction of growth of active tissues and organs. The role of ethylene during leaf development has been investigated physiologically using ethylene inhibitors and genetically using several ethylene-insensitive mutants such as ETR1, ER, EIN2 and EIN3 or transgenic plants that do not express the key enzymes of ethylene biosynthesis. All of these approaches showed that blocking ethylene biosynthesis or action could delay leaf development (Chao et al. 1997; Oh et al. 1999). An increase in leaf expansion has been observed in *Arabidopsis* and burst of ethylene was accompanied by an increased expression of ACC synthase gene 1, a gene suggested to be involved in the control of cell expansion (Rodrigues-Pousada et al. 1993). It appears that induction of ethylene biosynthesis is associated with leaf emergence or in the control of cell expansion. This is also supported from the studies on ethylene-insensitive mutants, which have a large rosette than the wild type

(Ecker 1995) resulting from cell enlargement (Hua et al. 1995). Lee and Reid (1997) found expansion in leaf area in sunflower with lower ethylene concentration. Ievinsh and Kreicbergs (1992), in cereal seedlings, reported that leaf emergence is associated with a peak of ethylene evolution.

Contradictory claims have been made on the effect of ethylene on photosynthesis with the use of ethylene-releasing compounds. An increase in the net photosynthetic rate (Grewal and Kolar 1990; Grewal et al. 1993; Khan et al. 2000) or decrease (Kays and Pallas 1980; Rajala and Peltonen-Sainio 2001) has been reported with the use of ethylene-releasing compounds. A correlation between ethylene-enhanced stomatal conductance and photosynthetic rate was found by Taylor and Gunderson (1986). A high concentration of CO_2 (>1%) acts as an antagonist of ethylene, but atmospheric CO_2 concentration is needed for the conversion of ACC oxidase to ethylene (Mattoo and White 1991). Thus, there is an interrelation between ethylene and CO_2 metabolism, and ethylene evolution controls the growth of plants. Bassi and Spencer (1982) showed an increase in ACC oxidase activity with an increase in CO_2 concentration. Dhawan et al. (1981), Kao and Yang (1982), and Grodzinski et al. (1982) also showed the inhibition of ethylene evolution resulting from a decrease in intercellular CO_2 concentration regulated photosynthesis. Similarly, Dong et al. (1992) showed completely abolished ACC oxidase activity in the absence of CO_2. ACC oxidase binds to Lys residue (Fernandez-Maculet et al. 1993) and results in carbamate formation (Ververides and Dilley 1994). Khan (2004a) has shown strong positive correlation of ACC synthase activity with photosynthetic rate and leaf area of mustard. The level of ACC synthase activity governed the photosynthetic rate and leaf area in two cultivars of mustard differing in photosynthetic activity. It is assumed that the structure of canopies in relation to leaf size and leaf area index is helpful in improving solar energy harvesting ability of leaves. Photosynthesis is responsive to leaf size, thus interception of solar radiation by leaves and its impact on photosynthesis.

Woodrow and Grodzinski (1989) suggested that CO_2 assimilation and growth might be altered by ethylene-related changes in leaf growth and total light interception. A lower concentration of ethylene increases the total leaf area while a high concentration reduces it (Mir 2003). Lee and Reid (1997) have also reported an increase in leaf area with low ethylene concentration. The studies with ethylene-insensitive mutants also implicate involvement of ethylene in leaf formation and expansion (Ecker 1995; Hua et al. 1995). Several explanations have been put forward for ethylene-mediated photosynthetic responses. Observations of Buhler et al. (1978), Grewal and Kolar (1990), Grewal et al. (1993) and Khan et al. (2000) have shown that the increase in photosynthesis with ethylene-releasing compounds was due to the increase in chlorophyll per unit of leaf area and greater light interception. Retaining high leaf area in ethephon (ethylene-releasing compound) treated plants helped in an increase in photosynthesis (Subrahmanyam and Rathore 1992; Khan et al. 2000).

The present work reports a study conducted to find the involvement of ACC synthase in the control of photosynthesis and growth. For that, two cultivars of mustard (*Brassica juncea* L. Czern & Coss.) having different photosynthetic capacities were used in the study. It was postulated that the cultivars would respond according to their capacities and produce optimal ethylene through ACC synthase to bring about maximal photosynthetic and growth responses. With this view, an ethylene modulator, ethephon (2-chloroethylphosphonic acid; an ethylene-releasing compound) was used to find a possible relationship of ethylene-mediated changes in photosynthesis with ACC synthase activity. To further strengthen the findings, the ACC synthase activity was quantified in the presence of modulators of ACC synthase activity, which was done by enhancing ACC synthase activity by spraying the plants with indole-3-acetic acid (IAA) or inhibiting the activity by spray of aminoethoxyvinylglycine (AVG). This was hypothesized that the high and low photosynthetic capacity cultivars would respond according to the IAA concentrations that might modify its capacity to produce optimal ethylene and maximal photosynthetic and plant growth response. Mustard was used as a model system because of its agricultural importance and wealth of available literature.

9.3.1 Ethylene in Mustard Cultivars Differing in Photosynthetic Capacity

An experiment on mustard (*Brassica juncea* L.) cvs. Varuna (high photosynthetic capacity) and RH30 (low photosynthetic capacity) showed that the ACC synthase activity and the ethylene pattern were similar to photosynthesis in the two cultivars. At different stages of plant development (30, 45, and 60 days after sowing), ACC synthase activity was higher in Varuna than RH30. Varuna also showed higher capacity of ethylene biosynthesis. The activity of ACC synthase was 74.4, 50.7, and 37.7% higher in Varuna than RH30 at 30, 45, and 60 days after sowing, respectively, whereas ethylene evolution was 78.6, 41.7, and 26.4% higher in Varuna than in RH30 at these growth stages. At all sampling times, photosynthetic rate in Varuna was higher showing 71.6, 47.1, and 29.9% increase over RH30. Varuna also had higher stomatal conductance and carbonic anhydrase activity. Leaf area and plant dry mass followed the pattern of rate of photosynthesis and ethylene in the cultivars.

The activity of ACC synthase was found to be correlated to the photosynthetic rate and leaf area in both the cultivars (Fig. 9.2). A strong positive correlation of ACC synthase activity with photosynthetic rate ($r^2=0.983$) and leaf area ($r^2=0.976$) was observed in Varuna. Also, in RH30 a positive relationship was found. In RH30, a correlation of ACC synthase activity with photosynthetic rate ($r^2=0.991$) and with leaf area ($r^2=0.976$) was noted (Khan 2004a).

The higher photosynthetic rate of Varuna was manifestation of ethylene-induced variation in stomatal and mesophyll effects. This caused greater influx of CO_2 increasing its fixation showing higher values for photosynthetic

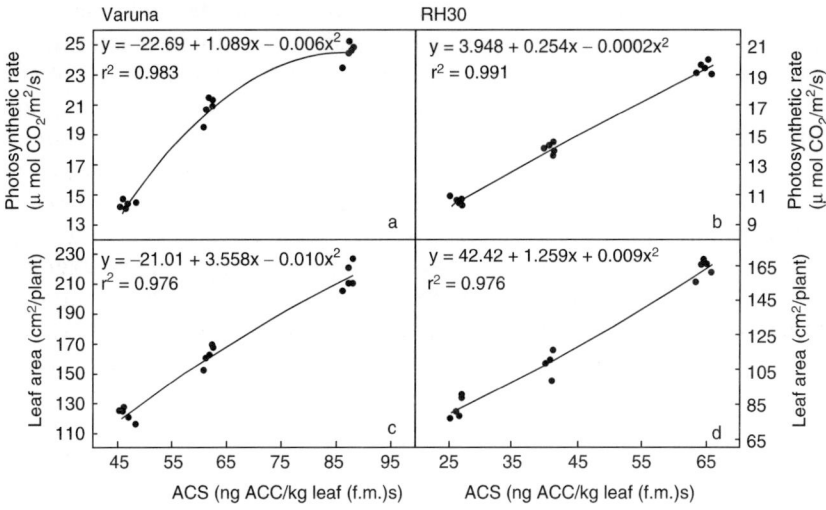

Fig. 9.2. Relationship of 1-aminocyclopropane carboxylic acid synthase (ACS) with net photosynthetic rate P_N, (**a, b**) and leaf area (**c, d**) of high photosynthetic capacity cultivar, Varuna (**a, c**) and low photosynthetic capacity cultivar, RH30 (**b, d**) of mustard (*Brassica juncea*) (Khan 2004a). With kind permission of Springer Science and Business Media

rate. The higher leaf area retained by the cultivar Varuna led to have an efficient light interception and thus photosynthesis and dry mass. The correlation studies between ACC synthase activity and photosynthesis and leaf area also suggest a role for ACC synthase in photosynthesis and growth.

9.3.1.1 Effects of Ethylene Modulators

Further to substantiate our understanding of the involvement of ethylene in photosynthesis, the two cultivars, Varuna and RH30, were treated with 0, 0.75, 1.5 and 3.0 mM ethephon at 30 days after sowing. Since ethephon on hydrolysis releases ethylene and phosphorus, an equivalent amount of phosphorus present in 3.0 mM ethephon was given to all treatments including the control to nullify the effects of phosphorus. Both the cultivars Varuna and RH30 responded maximally to 1.5 mM ethephon. Ethephon application significantly affected ACC synthase activity and ethylene evolution and were greatest with 3.0 mM of ethephon. The low photosynthetic capacity cultivar RH30 was more responsive to ethephon than the high photosynthetic capacity cultivar Varuna. Application of 1.5 mM of ethephon increased ACC synthase activity by 10.0% in Varuna and 29.0% in RH30. The same treatment of ethephon increased ethylene by 52.6% in Varuna and 75.0% in RH30. The increase in ethylene with 1.5 mM of ethephon was associated with the increase in photosynthetic rate, stomatal conductance, and carbonic anhydrase activity.

Ethephon at 1.5 mM increased the photosynthetic characteristics maximally; increasing the photosynthetic rate by 31.8 and 41.8%, stomatal conductance by 15.0 and 17.1%, and carbonic anhydrase activity by 84.6 and 71.4% in Varuna and RH30, respectively. The ratio of intercellular to ambient CO_2 concentration (Ci/Ca) was constant (Fig. 9.3) (Khan 2004b). An ethylene-induced increase in photosynthesis and stomatal conductance suggested that the differences in stomatal conductance contributed significantly to the variation in photosynthesis (Khan 2004b).

Leaf area and plant dry weight also responded similarly to photosynthetic characteristics; increasing to 1.5 mM ethephon treatment. A higher concentration of ethephon (3.0 mM) decreased the characteristics in both of the cultivars. It may be presumed that physiological concentration of ethylene in both cultivars was achieved with 1.5 mM of ethephon that triggered the physiological responses.

The favorable influence of ethephon on photosynthesis has been reported (Subrahmanyam and Rathore 1992; Khan et al. 2000). Increased stomatal conductance and carbonic anhydrase activity values in both high and low photosynthetic capacity cultivars, Varuna and RH30, respectively, reflected the stomatal conductance and mesophyll effects on photosynthesis. Mesophyll effects are characterized as a product of CO_2 binding capacity and the electron transport capacity. The carboxylation capacity determines the

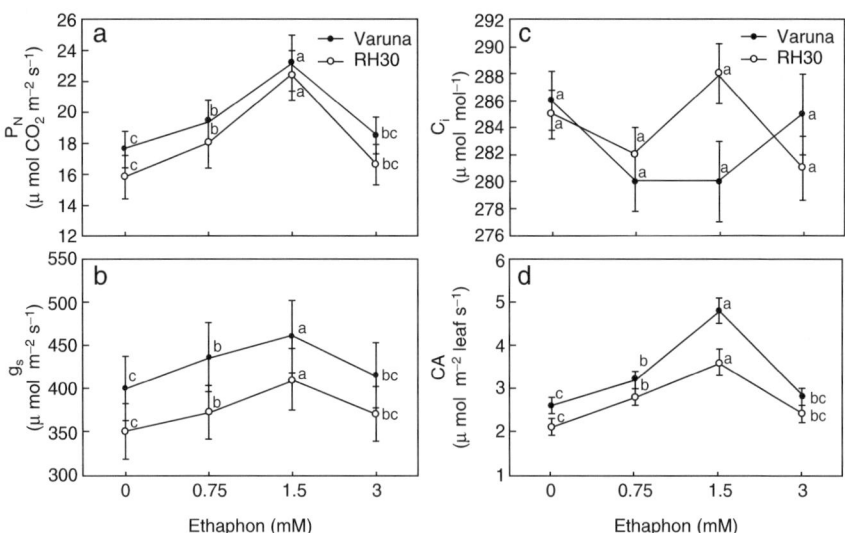

Fig. 9.3. Effect of different concentrations of ethephon on net photosynthetic rate, P_N **a** and stomatal conductance, g_s **b**, intercellular CO_2 concentration, C_i **c** and carbonic anhydrase (CA) activity **d** in high and low photosynthetic capacity cultivar, Varuna and RH30 of mustard (*Brassica juncea*). Each data point represent treatment mean ± SE. Values at each data point within the cultivar sharing the same letter are not significantly different at $P<0.05$ (Khan 2004b)

mesophyll effects (Pell et al. 1992; Eichelmann and Laisk 1999). An increase in carbonic anhydrase activity at the site of CO_2 fixation exhibited the enhanced carboxylation reaction (Badger and Price 1994; Moroney et al. 2001; Khan et al. 2004). The changes in stomatal conductance due to ethephon were a response to maintain stable intercellular CO_2 concentration under the given treatment. Thus, stomatal and mesophyll processes contributed to the increase in the photosynthetic rate in response to ethephon.

The ethephon-induced effects on photosynthetic parameters were mediated by ethylene evolved due to ethephon treatment. Taylor and Gunderson (1986) showed a relationship between ethylene-enhanced stomatal conductance and ethylene-enhanced photosynthesis. A higher concentration of ethephon (3.0 mM) decreased the photosynthesis and stomatal conductance. Such a condition of inhibition of photosynthesis by an ethylene-releasing compound has been observed by Kays and Pallas (1980) and Rajala and Peltonen-Sainio (2001). In all these studies, ethylene has been attributed to the changes in photosynthesis due to its effects on stomatal conductance. Mattoo and White (1991) reported that ethylene affected CO_2 assimilation and the plant responded depending on the tissue concentration. On the similar lines, Dhawan et al. (1981), Kao and Yang (1982) and Grodzinski et al. (1982) reasoned that a decrease in CO_2-regulated photosynthesis was related to ethylene evolution. In the present study, low photosynthetic capacity cultivar RH30 responded more to ethephon than the high photosynthetic capacity cultivar Varuna. In control plants, lower ethylene evolution in RH30 (due to low ACC synthase activity) than Varuna was responsible for the lower photosynthetic rate. As the ethylene evolution increased with ethephon application, the capacity of RH30 for photosynthesis also increased, resulting in a higher percent increase in photosynthesis than the Varuna. An increase of 75% ethylene (29% ACC synthase activity) in RH30 due to 1.5 mM ethephon increased the photosynthetic rate by 41.8%, whereas a 52.6% increase in ethylene (10% ACC synthase activity) in Varuna due to the same treatment increased photosynthesis by 31.8%. However, a further increase of ACC synthase activity and ethylene evolution with 3.0 mM ethephon proved inhibitory for photosynthesis.

It appeared therefore possible that the threshold value for ethylene and ACC synthase activity with 1.5 mM ethephon was comparable to that which elicits the ethylene-mediated hormonal responses, which differ with the cultivars inherent capacity of physiological processes. Low and high concentration represent the two ends of an optimum curve, promoting at low concentration and inhibiting at high.

9.3.1.2 Effect of ACC Synthase Activity Modulators

Finally, the study was carried out to strengthen our information on the role of ethylene in photosynthesis and growth. An approach of utilization of chemicals to modify ethylene biosynthesis was considered. For that reason, indole-3 acetic acid (IAA) and aminoethoxyvinylglycine (AVG) were applied on foliage of Varuna and RH30 cultivars of mustard. The effects of IAA and AVG on ACC synthase activity and photosynthesis were found significant in both low and high photosynthetic capacity cultivars, RH30 and Varuna, respectively. A concentration of 10^{-4} and 10^{-5} M spray enhanced the characteristics maximally in Varuna and RH30, respectively. Spray of 10^{-4} M on Varuna increased ACC synthase activity by 18.8% in comparison to the control. Similarly, photosynthesis was increased by 41.5% in comparison to the control with 10^{-4} M IAA. In the cultivar RH30, the maximal enhancement of 16.6% in ACC synthase activity and 39.0% in photosynthesis was noted with 10^{-5} M IAA. The application of AVG on both the cultivars resulted in a decrease in ACC synthase activity, ethylene, photosynthesis, and growth (Khan 2005b).

An increase in ACC synthase activity due to IAA application led to enhanced ethylene evolution, which manifested in ethylene-induced variation in stomatal conductance. This caused a greater influx of CO_2 and its fixation. The higher activity of carbonic anhydrase activity also reflects the increased rates of CO_2 fixation and higher photosynthesis. A relationship of carbonic anhydrase activity with photosynthesis has been suggested (Khan 1994; Khan et al. 2004). A higher concentration of CO_2 is known to act as an antagonist of ethylene, but atmospheric concentration of CO_2 is needed for the conversion of ACC to ethylene (Mattoo and White 1991). Thus, CO_2 could promote or inhibit ethylene evolution depending upon the sensitivity of the tissue. There is an interrelation between ethylene and CO_2 metabolism, affecting photosynthesis. The differential response of Varuna and RH30 to IAA concentration was due to the difference in the sensitivity of these cultivars for ethylene biosynthesis. The physiological active concentration of ethylene was achieved by 10^{-4} M IAA in Varuna and with 10^{-5} M in RH30. This overall effect resulted in leaf growth and plant dry weight. The reasons for this were explained earlier.

The inhibitor of ACC synthase activity, AVG reduced the ethylene biosynthesis and consequently reduced ethylene and, therefore, photosynthesis was lesser than the control. The reduced ethylene with the inhibition of ACC synthase activity was expected as ACC synthase is the first rate-limiting enzyme of ethylene biosynthesis.

9.4 Ethylene in Photosynthesis and Growth of Defoliated Plants

The removal of leaves, partial or complete, is described as defoliation. Responses of plants to defoliation are of considerable economic importance (McNaughton 1979a, 1979b). It modulates assimilate balance to the reproductive sink. Defoliation has been reported to cause increase in growth and physiological characteristics of plants, including emergence of new leaves with modified assimilatory capacity (Ericsson et al. 1980; Foggo 1996; Bruening and Egli 1999; Collin et al. 2000). The growth of defoliated plants is influenced by the nutritional status of plants (Thornton et al. 1994; Hamilton et al. 1998; Skinner et al. 1999), time of defoliation (Ericsson et al. 1980) and genotype (Nugent and Wagner 1995). Regarding defoliation, one may assume that foliar losses are balanced by increased irradiance of the leaves remaining after defoliation and that nutrient allocation patterns of plants depend on the severity of defoliation and whether the defoliation is done on the top or bottom canopy.

Light interception by the green organs and efficiency of photosynthetic conversion of intercepted light in biomass has influence on the productivity of plants. With this view, work of defoliation in mustard (*Brassica juncea* L. Czern & Coss.) was started as the factors of light and nutrients have special relevance because the leaves produced in mustard are large in number, broader in size, and oblong, and cause overshadowing effect. A presence of a large number of leaves affects the interception of photosynthetically active radiation (Ballare et al. 1989; Grewal and Kolar 1990; Kruger et al. 1998). The shaded and unproductive leaves in *Brassica* become senescent and abscise prematurely. The rate of tissue senescence declines markedly if photosynthetic tissues are maintained in the light by preventing shading. This could be achieved by adopting strategies that lead to rapid loss of lower leaves in the canopy at early growth stages. Thus, once the physiological cost to the plant of maintaining these dying leaves is removed, the assimilate balance may be improved. Significantly higher seed productivity on the removal of lower leaves of *Brassica juncea* has been found (Khan and Ahsan 2000; Khan 2002, 2003, 2005a; Khan et al. 2002a). Partial defoliation not only increases the relative demand for photosynthates on the remaining leaves but also increases the photosynthetic efficiency by reducing the competition between leaves for mineral nutrients and possibly for specific hormonal factors (Wareing et al. 1968). Increased light intensity upon the remaining photosynthetic tissues after defoliation results in an immediate increase in the photosynthetic rate per unit of remaining leaf (Jameson 1963; Robson 1973). The morphological and physiological changes following defoliation are presumed to have been brought about by signals produced by plant hormones. Therefore, the renewed growth of a terminal bud after removal of mature leaves may be associated with the changes in plant hormones. Qi and Yan (2000) reported accumulation of ABA in cotton leaves following defoliation. Earlier, in mustard, it has been reported (Khan 2003) that removal of 50% lower shaded leaves resulted in the increase in plant growth associated with ethylene and auxin concentration. Defoliation induces the production of phenolics in protection mechanism. The increased concentration of phenolics increases IAA oxidase activity and decreases auxin concentration. The index of relationship among ACC synthase activity, ethylene, photosynthetic rate, and growth suggests that there exists a concomitant relationship among these characteristics (Fig. 9.4) (Lone 2004).

Moreover, comparison of stage of defoliation, 40 days (pre-flowering) or 60 days (post-flowering) suggests that the early removal of leaves greatly influences photosynthesis and growth and also ACC synthase activity and ethylene (Khan and Lone 2005). Leaf removal at an early stage of reduces the competition between organs for efficient utilization of light, water, and nutrients. Moreover, younger leaves produced after defoliation are photosynthetically more active than those attained at later stages of plants cycle. The plants defoliated at an early stage show higher stomatal conductance, carbonic anhydrase activity, and CO_2 assimilation. Higher photosynthetic and growth activity of plants at this early stage (pre-flowering) also coincide with the higher activity of ACC synthase and ethylene. It may be suggested that increased activity of ACC synthase and ethylene are attributed to the emergence of more new leaves at the early stage of development. A comparison of

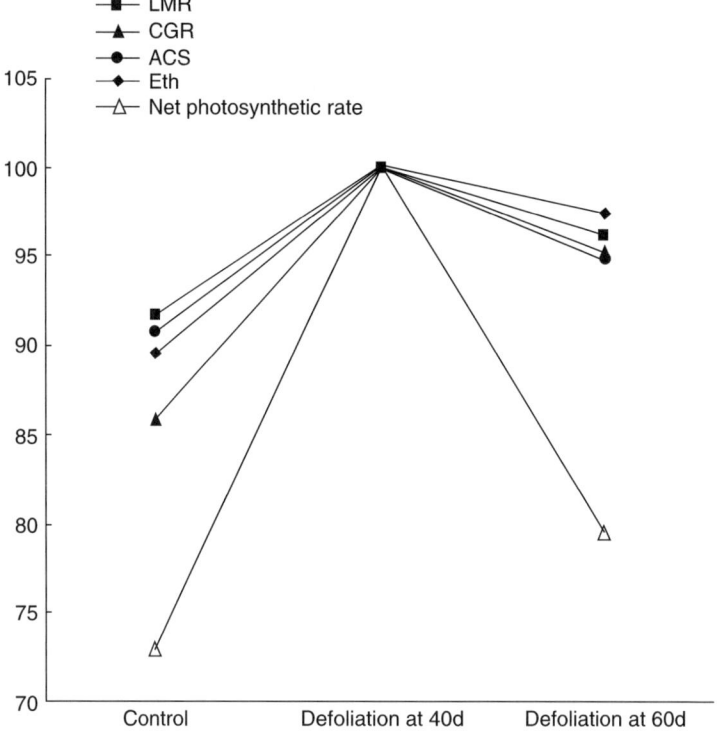

Fig. 9.4. Index of relationships among leaf mass ratio (LMR), crop growth rate (CGR), ACC synthase (ACS) activity, ethylene (Eth) evolution and net photosynthetic rate of mustard (*Brassica juncea*) defoliated at 40 or 60 days after sowing

50% upper or lower leaves suggested that upper leaves removal showed higher decrease in photosynthetic rate but greatest ethylene evolution. Application of auxins (IAA, IBA, and NAA) helped in the partial restoration of the decrease in photosynthesis and growth (Khan et al. 2002a). Ethylene evolution in this study was shown to be determinant for morphology and physiology of the plant. Another attempt to substantiate the hypothesis that ethylene has a role in photosynthesis and growth, ethylene application on intact plants and plants with 50% lower leaves removed were compared. Photosynthesis and growth were increased with removal of leaves. Ethephon at 200 µL L^{-1} increased the overall growth of plants in no-defoliation treatment, which was equivalent to defoliation plants treated with water spray. Ethephon spray on defoliated plants proved inhibitory. The ethylene concentration in 200 µL L^{-1} ethephon treatment on no-defoliation plants was equal to water spray on defoliated plants. The growth and photosynthesis changes were found to be correlative with the changes in the ethylene level. Ethephon at 200 µL L^{-1} increased the ethylene in no-defoliation plants that increased the characteristics maximally. However, in defoliation plants, such ethylene concentration was achieved in water spray. Ethephon applied on defoliation plants resulted in supra-optimal ethylene concentration that reduced the characteristics (Khan 2005a). Based on these studies, it may be said that among several intrinsic and extrinsic factors that control photosynthesis of plants, ethylene plays a prominent role. Ethylene-induced abscission is associated with the expression of polygalacturonase and ends β-1,4 glucan hydrolase in the vicinity of the distal abscission zone. However, it is also not be ruled out that ethylene may not act as a single developmental switch setting a cascade in motion, but may act as an inducer whose presence is required for long periods of time.

References

Abeles FB, Morgan PW, Saltveit ME Jr (1992) Ethylene in plant biology, 2nd edn. Academic, San Diego

Andrews TJ, Lorimer GH (1987) Rubisco. Structure, mechanism and prospects for improvement. In: Hatch MD, Boradiman NK (eds) A comprehensive treatise, vol. 10. Photosynthesis. Academic, New York, pp 131–218

Arteca RN (1997) Plant growth substances: principles and applications. CBS Publishers and Distributors, Delhi

Arteca RN, Dong CN (1981) Stimulation of photosynthesis by application of phytohormones to the root system of tomato plants. Photosynth Res 2:243–249

Badger RR, Price GD (1994) The role of carbonic anhydrase in photosynthesis. Annu Rev Plant Physiol Plant Mol Biol 45:369–393

Ballare CL, Scopel AL, Sanchez RA (1989) Photomodulation of axis extension in sparse canopies: role of stem in the perception of light quality signals of stand density. Plant Physiol 89:1324–1330

Bassi PK, Spencer MS (1982) Effect of carbon dioxide and light on ethylene production in intact sunflower plants. Plant Physiol 69:1222–1225

Bode J, Wild A (1984) The influence of 2-chloroethyl trimethyl ammonium chloride (CCC) on growth and photosynthetic metabolism of young wheat plants (*Triticum aestivum* L.). J Plant Physiol 116:435–446

Bradford KJ, Trewavas AJ (1994) Sensitivity threshold and variable time scales in plant hormone action. Plant Physiol 105:1029–1036

Bruening WP, Egli DB (1999) Relationship between photosynthesis and seed number of phloem isolated nodes in soybean. Crop Sci 39:1769–1775

Buhler B, Drumm H, Mohr H (1978) Investigations on the role of ethylene in phytochrome-mediated photomorphogenesis. II. Enzyme levels and chlorophyll synthesis. Planta (Berlin) 142:119–122

Chao Q, Rothenberg M, Solano R, Roman G, Terzaghi W, Ecker J (1997) Activation of the ethylene gas response pathway in *Arabidopsis* by the nuclear protein: ethylene insensitivities and related proteins. Cell 89:1133–1144

Child RD, Arnold G, Hislop EC, Huband NDS, Stinchcanbe GR (1985) Effects of some experimental triazole retardants on yield of oil seed rape. Proc Brit Crop Protection Conf, Weeds 2:561–567

Ciardi J, Klee H (2001) Regulation of ethylene mediated responses at the level of the receptor. Ann Bot 88:813–822

Collin P, Epron D, Alaoui-Sosse B, Badot PM (2000) Growth responses of common ash seedlings (*Fraximus excelsior* L.) to total and partial defoliation. Ann Bot 85:317–323

Dhawan KR, Basu PK, Spencer MS (1981) Effects of carbon dioxide on ethylene production and action in intact sunflower plants. Plant Physiol 68:831

Dijkstra P, Reegan H, Kuiper PJC (1990) Relation between relative growth rate, endogenous gibberellins and the response to applied gibberellic acid on *Plantago ovata*. Physiol Plant 79:629–634

Dong JG, Fernandez-Maculet JC, Yang SF (1992) Purification and characterization of 1-aminocyclopropane-1-carboxylate from apple fruit. Proc Natl Acad Sci USA 89:9789–9793

Ecker JR (1995) The ethylene signal transduction pathway in plants. Science 268:667–675

Eichelmann H, Laisk A (1999) Ribulose-1,5-bisphosphate carboxylase/oxygenase content, assimilatory change and mesophyll conductance in leaves. Plant Physiol 119:179–189

Ericsson A, Larson S, Tenow O (1980) Effects of early and late season defoliation on growth and carbohydrate dynamics in scot pine. J Appl Ecol 17:747–769

Fernandez-Maculet JC, Dong JG, Yang SF (1993) Activation of 1-aminocyclopropane-1-carboxylate oxidase by carbon dioxide. Biochem Biophys Res Commun 193:1168–1173

Foggo A (1996) Long- and short-term changes in plant growth following simulated herbivory: adaptive responses to damage. Ecol Entomol 21:198–202

Foroutan-Pour K, Ma BL, Smith DL (1997) Protein accumulation potential in barley seeds as affected by soil- and peduncle-applied N and peduncle-applied plant growth regulators. Physiol Plant 100:190–201

Grewal HS, Kolar JS (1990) Response of *Brassica juncea* to chlorocholine chloride and ethrel sprays in association with nitrogen application. J Agric Sci 114:87–91

Grewal HS, Kolar JS, Cheema SS, Singh G (1993) Studies on the use of growth regulators in relation to nitrogen for enhancing sink capacity and yield of gobhi-sarson (*Brassica napus*). Indian J Plant Physiol 36:1–4

Grodzinski B, Boesel L, Horton RF (1982) Ethylene release from leaves of *Xanthium strimarium* L. and *Zea mays* L. J Exp Bot 33:344–354

Grossmann K, Heusen H (2001) Ethylene triggered abscisic acid. A principal in plant growth regulation? Physiol Plant 113:9–14

Hamilton WE III, Giovannini MS, Moses SA, Coleman JS, McNaughton SJ (1998) Biomass and mineral element responses of a Serengeti short-grass species to nitrogen supply and defoliation: compensation requires a critical N. Oecologia 116:407–418

Heide OM, Bush MG, Evans LT (1985) Interaction of photoperiod and gibberellin on growth and photosynthesis of high latitude *Poa pratensis*. Physiol Plant 65:135–145

Hua J, Chang C, Sun Q, Meyerowitz EM (1995) Ethylene insensitivity conferred by *Arabidopsis* ERS gene. Science 269:1712–1714

Hussain A, Black CR, Taylor IB, Roberts JA (1999) Soil compaction: a role of ethylene in regulating leaf growth. Plant Physiol 121:1227–1237

Ivenish G, Kreicbergs O (1992) Endogenous rhythmicity of ethylene production in growing intact cereal seedlings. Plant Physiol 100:1389–1391

Jameson DA (1963) Responses of individual plants to harvesting. Bot Rev 29:532–594

Kao CH, Yang SF (1982) Light inhibition of the conversion of 1-aminocyclopropane-1-carboxylic acid to ethylene in leaves is mediated through carbon dioxide. Planta 155:261–266

Kasele IN, Shanahan JF, Nielsen DC (1995) Impact of growth retardants on corn leaf morphology and gas exchange traits. Crop Sci 35:190–194

Kays SJ, Pallas JE Jr (1980) Inhibition of photosynthesis by ethylene. Nature 385:51–52

Kende H (1993) Ethylene biosynthesis. Annu Rev Plant Physiol Mol Biol 44:283–307

Kende H, Zeevaart JAD (1997) The five classical plant hormones. Plant Cell 9:1197–1210

Khan NA (1994) Variation in carbonic anhydrase activity and its relationship with photosynthesis and dry mass of mustard. Photosynthetica 30:317–320

Khan NA (1996a) Effect of gibberellic acid on carbonic anhydrase, photosynthesis, growth and yield of mustard. Biol Plant 38:145–147

Khan NA (1996b) Response of mustard to ethrel spray and basal and foliar application of nitrogen. J Agron Crop Sci 176:331–334

Khan NA (2002) Activities of carbonic anhydrase and ribulose-1,5-bisphosphate carboxylase, and plant dry mass accumulation in *Brassica juncea* following defoliation. Photosynthetica 40:633–634

Khan NA (2003) Changes in photosynthetic biomass accumulation, auxin and ethylene level following defoliation in *Brassica juncea*. J Food Agric Environ 1:125–128

Khan NA (2004a) Activity of 1-aminocyclopropane carboxylic acid synthase in two mustard (*Brassica juncea* L.) cultivars differing in photosynthetic capacity. Photosynthetica 42:477–480

Khan NA (2004b) An evaluation of the effects of exogenous ethephon, an ethylene releasing compound, on photosynthesis of mustard (*Brassica juncea*) cultivars that differ in photosynthetic capacity. BMC Plant Biol 4:21

Khan NA (2005a) The influence of exogenous ethylene on growth and photosynthesis of mustard (*Brassica juncea*) following defoliation. Sci Hort 105:499–505

Khan NA (2005b) Involvement of 1-aminocyclopropane carboxylic acid synthase in photosynthesis and growth of mustard cultivars with different photosynthetic capacities. Physiol Mol Biol Plants 11:333–337

Khan NA, Ahsan N (2000) Evaluation of yield potential of defoliated mustard cultivars. Tests Agrochem Cult 21:33–34

Khan NA, Lone PM (2005) Effects of early and late season defoliation on photosynthesis, growth and yield of mustard (*Brassica juncea* L.). Braz J Plant Physiol 17:181–186

Khan NA, Ansari HR, Mobin M (1996) Effect of gibberellic acid and nitrogen on carbonic anhydrase activity and mustard biomass. Biol Plant 38:601–603

Khan NA, Lone NA, Samiullah (2000) Response of mustard (*Brassica juncea* L.) to applied nitrogen with or without ethrel spray under non-irrigated conditions. J Agron Crop Sci 183:1–4

Khan NA, Khan M, Ansari HR, Samiullah (2002a) Auxin and defoliation effects on photosynthesis and ethylene evolution in mustard. Sci Hort 96:43–51

Khan NA, Mir R, Javid S, Khan M, Samiullah (2002b) Effects of gibberellic acid spray on nitrogen yield efficiency of mustard grown with different nitrogen levels. Plant Growth Regul 38:243–247

Khan NA, Singh S, Khan M, Samiullah (2003) Interactive effect of nitrogen and plant growth regulators on biomass partitioning and seed yield: mediation by ethylene. Brassica 5:64–71

Khan NA, Javid S, Samiullah (2004) Physiological role of carbonic anhydrase in CO_2 fixation and carbon partitioning. Physiol Mol Biol Plants 10:153–166

Kieber JJ, Rothenberg M, Roman G, FieldmannKA, Ecker JR (1993) CTRI: a negative regulator of the ethylene response pathway in *Arabidopsis*: encodes a number of the Raf family of protein kinases. Cell 72:427–441

Kruger EL, Volin JC, Lindroth RL (1998) Influence of atmospheric CO_2 enrichment on the responses of sugar maple and trembling aspen to defoliation. New Phytol 140:85–94

Lee SH, Reid DM (1997) The role of endogenous ethylene in the expansion of *Helianthus annus* leaves. Can J Bot 75:501–508

Liu HS, Peng WB, Meng FT, Yuan JP, Wang DQ (1993) Effects of S-3307 on morphological and some physiological characteristics of wheat seedlings. Plant Physiol Commun 29:354–355

Lone PM (2004) Growth and metabolism of mustard (*Brassica juncea*) following defoliation and nitrogen treatment. PhD Thesis. Aligarh Muslim University, Aligarh, India

Lorimer GH, Miziorka HM (1980) Carbamate formation of the e-amino group of Lysyl residue as a basis for the activation of ribulose bisphosphate carboxylase by CO_2 and Mg^{+2}. Biochemistry 19:5321–5328

Makeev AV, Krendeleva TE, Mokronsov AT (1992) Photosynthesis and abscisic acid. Soviet Plant Physiol 39:118–126

Mathooko FM (1996) Regulation of ethylene biosynthesis in higher plants by carbon dioxide. Postharvest Biol Technol 7:1–26

Mattoo AK, White WB (1991) Regulation of ethylene biosynthesis. In: Mattoo AK, Suttle JC (eds) The plant hormone ethylene. CRC Press, Boca Raton, pp 21–42

McNaughton SJ (1979a) Grazing as an optimization process: grassungulate relationships in the Serengeti. Am Naturalist 113:691–703

McNaughton SJ (1979b) Grassland-herbivore dynamics. In: Sinclair ARE, Norton-Griffiths M (eds) Serengeti: studies of ecosystem dynamics in a tropical savanna. Univ Chicago Press, Chicago, pp 46–81

Mir MR (2003) Physiological significance of ethrel (2-chloroethyl phosphonic acid) on growth and metabolism of mustard under irrigated and non-irrigated conditions. PhD Thesis, Aligarh Muslim University, Aligarh, India

Moroney JV, Bartlett SG, Samuelsson G (2001) Carbonic anhydrase in plants and algae. Plant Cell Environ 24:141–153

Nagel OW, Lambers H (2002) Changes in acquisition and partitioning of carbon and nitrogen in the gibberellin deficient mutants A70 and W335 of tomato (*Solanum lycopersicumi*). Plant Cell Environ 25:883–891

Neljubow DN (1901) Über die horizontale Nutation der Stengel von Pisum sativum und einiger anderen Pflanzen. Beih Bot Centralbl 10:128–139

Nugent SP, Wagner MR (1995) Clone and leaf position effects on *Populus* defoliation by leaf cutting bees (*Hymenoptera megachilidae*). For Ecol Manage 42:485–467

Oh SA, Park JH, Lee GI, Pack KH, Park SK, Nam HG (1999) Identification of three genetic loci controlling leaf senescence in *Arabidopsis thaliana*. Plant J 12:527–533

Pell EJ, Eckardt NA, Eryedi AJ (1992) Timing of ozone stress and resulting status of ribulose bisphosphate carboxylase oxygenase and associated net photosynthesis. New Phytol 120:397–405

Qi XL, Yan SL (2000) Dynamics of plant endogenous hormone of cotton leaves infected by *Verticilium dahliae*. Acta Gossypii Sinica 12:310–312

Rajala A, Peltonen-Sainio P (2001) Plant growth regulator effects on spring cereal root and shoot growth. Agron J 93:936–943

Rakitina TY, Vlasov PV, Jalilova FK, Kafeli V (1994) Abscisic acid and ethylene in mutants of *Arabidopsis thaliana* differing in their resistance to ultraviolet (UV-B) radiation stress. Russian J Plant Physiol 41:599–603

Raskin I, Kende H (1984) The role of gibberellin in the growth response of submerged deep water rice. Plant Physiol 76:947–950

Ridge I, Osborne DJ (1971) Role of peroxidase when proline rich protein in plant cell walls is increased by ethylene. Nature New Biol 229:205–208

Robson MJ (1973) The growth and development of simulated swards of perennial ryegrass. I. Leaf growth and dry weight changes as related to the ceiling yield of a seedling sward. Ann Bot 37:487–500

Rodrigues-Pousida RA, de Rycke R, Dedondar A, van Caeneghem W, Engler E, van Montagu M, van der Straeten D (1993) The Arabidopsis 1-amino-cyclopropane-1-carboxylate synthase gene 1 is expressed during early development. Plant Cell 5:897–911

Sairam RK, Deshmukh PS, Shukla DS (1991) Influence of chloromequate chloride on photosynthesis and nitrate assimilation in wheat genotypes under water stress. Indian J Plant Physiol 34:222–227

Sharp RE, Le Noble ME (2002) ABA, ethylene and the control of shoot and root growth under water stress. J Exp Bot 531:33–37

Shibaoka H (1994) Plant hormone induced changes in the orientation of cortical microtubules: alterations in the crosslinking between microtubules and plasma membrane. Annu Rev Plant Physiol 45:527–544

Skinner RH, Morgan JA, Hanson JD (1999) Carbon and nitrogen reserve remobilization following defoliation: nitrogen and elevated CO_2 effects. Crop Sci 39:1749–1756

Smalle J, van der Straeten D (1997) Ethylene and vegetative development. Physiol Plant 100:593–605

Spollen WG, LeNoble ME, Samuels TD, Bernastein N, Sharp RE (2000) Abscisic acid accumulation maintains maize primary root elongation at low water potential by restricting ethylene production. Plant Physiol 122:967–976

Subrahmanyam D, Rathore VS (1992) Influence of ethylene on carbon-14 labeled carbondioxide assimilation and partitioning in mustard. Plant Physiol Biochem 30:81–86

Tan ZY, Thimann KV (1989) The role of carbon ethylene. Physiol Plant 75:13–19

Taylor GE Jr, Gunderson CA (1986) The response of foliar gas exchange to exogenously applied ethylene. Plant Physiol 82:653–657

Thetford M, Warren SL, Blazick FA, Thomas JF (1995) Response of *Forsynthia-x intermedia Spectabilis* to unicongzole. 2. Leaf and stem anatomy, chlorophyll and photosynthesis. J Am Soc Hort Sci 120:983–988

Thornton B, Millard P, Duff EI (1994) Effects of nitrogen supply on the source of nitrogen used for regrowth of laminae after defoliation of four grass species. New Phytol 128:615–620

Trewavas AJ (1983) Is plant development regulated by changes in the concentration of growth substances or by changes in the sensitivity to growth substances? Trends Biochem Ser 8:354–357

Urwiller MJ, Stutte CA (1986) Influence of ethephon on soybean reproduction development. Crop Sci 26:976–979

Ververides P, Dilley DR (1994) Mechanism studies of CO_2 activation for a lysyl residue involvement. Plant Physiol [Suppl] 105:33

Wang KLC, Li H, Ecker JR (2002) Ethylene biosynthesis and signaling networks. Plant Cell 14:1–38

Wareing PF, Khalifa MM, Treharne KJ (1968) Rate-limiting processes in photosynthesis at saturating light intensities. Nature 220:453–457

Warner HL, Leopold AC (1969) Ethylene evolution from 2-chloroethylphosphonic acid. Plant Physiol 44:156–158

Woodrow L, Grodzinski B (1989) An evaluation of the effects of ethylene on carbon assimilation in *Lycopersicon esculentum* Mill. J Exp Bot 40:361–368

Yang Q, Yang JX, Hu YW (1994) Effects of S-3307 on some physiological characteristics of rape seedlings. Plant Physiol Comm 33:182–185

Yuan L, Xu DQ (2001) Stimulation effect of gibberellic acid short-term treatment on leaf photosynthesis related to the increase in Rubisco content in broad bean and soybean. Photosynthesis Res 68:39–47

Zerbe R, Wild A (1981) The effect of indole-3-acetic acid on photosynthetic apparatus of *Sinapis alba*. Photosynthesis Res 1:71–81

Index

ABA 41, 44, 82
Abiotic 81
Abscissic acid 187
Abscission 23, 54, 99, 197
ACC 52, 72, 82
ACC deaminase 6, 124, 170
ACC oxidase, ACO 17, 53, 82, 154, 155
ACC synthase, ACS 17, 82, 154, 190
Acclimation 81, 108
Acetylene 2
ACS genes 154
Acetylsalicylic acid, AA 85
Adaptation 102
Adventitious 70
Aeration 94
Alkenes 13
Anoxic 94
Antagonist 27
Antagonistic 54
Antagonistic potency 12
Antibiotic 52
Antifreeze activity 110
Antioxidant 59
Aminooxyacetic acid, AOA 55
Apical hook 69
Apoplast 110
Arabidopsis 135, 136, 138, 140–143, 145–147
Ascorbate peroxidase 60
Asr1 165
ASR protein 160
AtLecRK2 101, 102
ATPase 16
Autocatalysis 86, 87, 88, 89
Autoinhibition 86, 87
Autoregulation 129
Auxin 54, 99, 135, 138, 144–147
AVG 43, 55, 70

Binding site 13
Biological activity 14
Blue light 44
Bromodeoxyuridine 76

Calcium spiking 125
Carbon monoxide 2
Carbonic anhydrase 190
Carboxylation 192
Carotenoid 63
Catalase 59
Cellulase 172
Chilling 83, 95–97, 99, 101, 102, 109, 110
Chlorophyll 52, 89
Chlorophyll degradation 22
Cis-elements 174
Climacteric 97
Climacteric fruits 151
Cnr 159
Compacted soil 106, 107
Constitutive 97
Copper 3
Copper transport 16
Crack-entry 126
β-cyanolalanine synthase, CAS 82, 83, 106
Cyclic alkenes 7
CTR1 99, 100, 102
Cyclopropene 10
Cyclopropene derivatives 12

Decarboxylated SAM, dSAM 82, 83
Defoliation 194
Dehydration 90
Dehydroascorbate reductase 60
DELLA proteins 40
Dessication 89

det1 gene 174
Determinate nodule 121
Diazocyclopentadiene 9
Double bond 10
Drought stress 44, 45
Cytokinin 141
3,3-Dimethylcyclopropene 11
^{14}C- Ethylene 4

E4 162, 175
E8 162
E8 promoter 171
EILs 100
EIN2 100, 102
EIN3 100, 156
EIN4 99, 101
Electron transport capacity 192
EPA 24
Epinasty 103, 108, 109
EREBPs 100
ERF1 100, 156
ERS1 99, 101, 102
ERS2 99, 101
EthylBloc 21
Ethylene antagonist 7
Ethylene binding 4
Ethylene perception 15
Ethylene receptor 3
Ethylene responsive genes 151
Ethylene signal transduction 36
Ethylene-binding domain 8
Ethylene-insensitive plants 36, 39, 45
ETR1 5, 36, 39, 42
ETR2 99, 100, 102
Expansins 40, 161
Explant 71

Flooding 83, 93, 94, 101, 103, 109
Floriculture 57
Flower senescence 21
Flower sleepiness 22
Freezing 95
Fruit firmness 20
Fruit specific expression 175

α-Galactosidase 172
1-(γ-L-glutamylamino)ACC, GACC 82, 83

Galactosidase 161
Gas exchange 187
GCC box 160
Gibberellic acid 186
Gibberellins 61
β-glucuronidase, GUS 86–88, 91, 93, 97, 106
Glutathione 60
Glutathione S-tranferase 64
Gravitropism 135, 137, 138, 140, 141, 143–147
Green fluorescence protein, GFP 86, 97
Growth inhibition 38, 39
Growth promotion 39, 42, 43

Heterodimeric isoenzymes 111
Histidine kinase 99, 101
Hormone 1
Horticulture 57
Hydrolytic 55
Hydrophobic interaction 13
Hydroponic roots 126
Hypocotyl 70, 136, 137, 140–146
Hypoxia 83, 89, 93, 95, 101, 109

IAA 99, 100, 109
IBA 74
Indeterminate nodule 120
Infection thread abortion 123
Inflorescence stalks 135, 136, 138, 140–143
Isogenes 88

Jasmonate, J 83, 85
Jasmonic acid, JA 18, 85, 100, 101, 103–105, 109

Lateral root base (LRB) nodulation 120
Leaf area 37–40, 43
Leaf area index 186
Leaf expansion 37, 38, 39, 42, 43
Legume 119
Ligand 3
Ligand rearrangement 15
Light 136–138, 140, 141, 143, 147
Light interception 189, 195
Lipid 51
Long-distance communication 129

Index

MADS box protein 159
MADS transcription factor 159
MAP kinase cascade 156
1-methylcyclopropane, MCP 56, 84, 87, 89, 96, 99, 102, 106, 109, 156
Membrane-targeting agents 14
Mesophyll effects 190
Methionine, Met 69, 82, 83
Methyl jasmonate, MJ 85, 104, 105
1-Methylcyclopropene 11
3-Methylcyclopropene 11
Microtubili 40
Microtubule 188
Mutant 16

Natural antagonist 26
Necrosis 53
Never ripe (Nr) 155
Nitric acid 75
Nodulation factor 119
Nodulin gene 121
Nodule number 122
Nodule plasticity 127
Nodule primordium positioning 128
Non climacteric fruits 152
Norbornadiene, NBD 6, 85
Nutrients 44
N-malonylACC, MACC 82, 83, 91, 95
NTHK1 101, 102

Organogenesis 72
Osmotic potential 92
Osmotic stress 92, 93, 98, 102, 108
Oxidative stress 98, 99, 106, 108
Oxygen 25
Ozone 83, 97, 98, 101–106

Pea 135, 136, 138–140, 143–144
Peroxidase 17, 188
Phenolics 195
Phenotypic 98, 100, 107, 109
Phosphorus 191
Phosphorylation 88
Photoacoustic 97
Photosynthesis 35–38, 41–45, 51, 185
Photosynthetically active radiation 185
Plant growth 19
Plant growth regulators 185
Pollination 18

Polyamines, PA 82, 83, 100, 108
Polyethylene glycol 93
Post-climacteric 153
Post-harvest 24, 81, 82, 84
Pre-climacteric 153
Programmed cell death, PCD 75, 100, 103
Propylene 2
Protein turnover 83
Protoxylem poles 128
Pseudonodule 121
Putative 101

Rab11A GTPase 173
Radiation 195
Reactive oxygen species, ROS 59, 83, 85, 91, 93, 100, 103, 104, 106, 108, 111
Relative growth rate 35, 38, 43, 45
Reporter genes 86–88, 91, 93, 97
Respiration 18
Re-watering 90, 109
Rhizobiaceae 119
Rhizobial infection 122
Rhizobitoxine 124
Rhizogenesis 69
Ring strain 9
Ripening 151
Ripening control 20
Ripening inhibitor (rin) 155
Root hair curling (RHC) 120
Root hair deformation 125
Rubisco 38, 41

S-adenosylmethionine decarboxylase, SAMDC 82, 83, 108, 167
S-adenosylmethionine, SAM 82, 83, 108
Salicylic acid, SA 71, 83, 91, 100, 104, 105
Salinity 83, 91–93, 101–103, 107
Salt stress 44, 45
Salt tolerance 108, 109
SAM synthetase, SAMS 82, 83, 156
Senescence 35, 39, 41, 43, 44, 51
Sesbania rostrata 125
Shade avoidance 44
Sickle mutant 123
Signal transduction 4
Silver nitrate 87, 110
Silver thiosulphate, STS 56, 87, 99, 107

SLA see specific leaf area
SmartFresh 21
Specific leaf area 37, 38, 43
Stomatal conductance 41, 43, 92, 189
Stress 19
Stress protectent 108
Submergence 76, 93–95, 101, 109
Sugar sensing 41, 42
Synergistic 61

Terpenes 8
Thick short roots (Tsr) 122
Thigmomorphogenesis 103
Tolerance 103

Tomato 136, 137, 140, 141, 143, 145
Toxic ions 91
Trans-cyclooctene 7
Treatment conditions 25
Triple response 23, 38

Water deficit 89
water potential, (Ø) 89–91, 107
Waterlogging 93, 94, 109, 126
Wounding 84

Yeast 5
Yellowing 63